Wolbachia: A Bug's Life in another Bug

Issues in Infectious Diseases

Vol. 5

Series Editors

Heinz Zeichhardt Berlin
Brian W. J. Mahy Atlanta, Ga.

Wolbachia: A Bug's Life in another Bug

Volume Editors

Achim Hoerauf Bonn
Ramakrishna U. Rao St. Louis, Mo.

25 figures, 7 in color, and 1 table, 2007

KARGER

Basel · Freiburg · Paris · London · New York ·
Bangalore · Bangkok · Singapore · Tokyo · Sydney

Issues in Infectious Diseases

Achim Hoerauf
Institute for Medical Microbiology,
Immunology and Parasitology
University Clinic Bonn
53105 Bonn, Germany

Ramakrishna U. Rao
Department of Internal Medicine
Infectious Diseases Division
Washington University School of Medicine
St. Louis, Mo. 63110 USA

Library of Congress Cataloging-in-Publication Data

Wolbachia : a bug's life in another bug / volume editors, Achim Hoerauf,
Ramakrishna U. Rao.
p. ; cm. – (Issues in infectious diseases, ISSN 1660-1890 ; v. 5)
Includes bibliographical references and indexes.
ISBN 978-3-8055-8180-6 (hard cover : alk. paper)
1. Wolbachia. 2. Filariasis. 3. Filarial infections. 4. Nematoda as
carriers of disease. I. Hoerauf, Achim. II. Rao, Ramakrishna U. III.
Series.
[DNLM: 1. Wolbachia. 2. Filarioidea–microbiology. 3.
Filarioidea–parasitology. 4. Host-Parasite Relations. 5. Nematode
Infections–parasitology. QW 150 W848 2007]
QR353.5.R5W65 2007
614.5′552–dc22

2007007801

Bibliographic Indices. This publication is listed in bibliographic services, including Current Contents® and Index Medicus.

www.karger.com
Printed in Switzerland on acid-free paper by Reinhardt Druck, Basel
ISSN 1660–1890
ISBN 978–3–8055–8180–6

Contents

Foreword

This comprehensive publication is intended for readers with teaching or research interests in microbiology, entomology, infectious diseases, genetics, tropical medicine, and clinical research.

Worldwide, approximately 120 million people are infected by filarial nematode parasites. Transmitted to humans through mosquitoes and black flies, the majority of the disease-causing nematodes are hosts to the *Wolbachia* bacteria. These nematodes cause the often devastating diseases elephantiasis and onchocerciasis, commonly referred to as filariasis. Moreover, heartworm disease, caused by another *Wolbachia*-containing nematode, is another mosquito-borne disease that has significant importance in the veterinary field.

This textbook in the infectious disease series intends to comprise a reference for researchers in the field of drug discovery, as antibiotics and antiwolbachia formulations will aid in eliminating disease transmission and pathogenesis. Entomologists may be interested in this work since *Wolbachia* infections in some arthropods have been known to the scientific community for several years, and the biological and biochemical relationships between *Wolbachia* and their insect hosts have been fascinating the insect research community. Much progress has been made studying insect *Wolbachia* genes in influencing insect populations and behavior. Successful manipulations with *Wolbachia* transgenes in mosquito vectors may eventually lead to control of the vector-borne diseases in humans and animals. In early 2007, more than 700 research papers indexed in PubMed were associated with *Wolbachia* and most of them were related to its role in arthropods. Since the identification of *Wolbachia* endobacteria in filarial nematodes, the number of research papers on this subject has steadily increased with some very interesting findings and enhanced the vision of eliminating the dreadful disease 'filariasis' sometime in the near future. In 2005, three decades after the discovery of *Wolbachia* in filarial nematodes, the genome sequencing of nematode *Wolbachia* was completed. Detailed comparisons to insect *Wolbachia* genome were performed and made available to the public, leading to new insights in their relationships. This publication features a mixture of internationally recognized leaders in infectious disease research and insect biology. Their interesting perspectives on *Wolbachia's* genome, evolution, symbiosis, biology, pathogenicity as well as its potential as a drug target are some of the highlights of this book.

Chapter 1, written by one of the pioneers in identifying the bacteria in filarial nematodes, addresses the historical perspectives and highlights what we have learned about *Wolbachia* then and now. Chapter 2 details the evolution and phylogeny of the filarial *Wolbachia* lineage in comparison with *Wolbachia* found in other organisms. Chapter 3 provides a review on *Wolbachia* as a target for chemotherapy and its current status in human clinical trials. Chapter 4 highlights and updates the current understanding of the *Wolbachia* genome and the mining of *Wolbachia* genes in the genome, which is useful to identify the bacterial relationships with their nematode hosts. Chapter 5 expands our understanding of the *Wolbachia* genome of filarial nematodes in comparison with insect *Wolbachia* genome including recent studies involving the lateral gene transfer between bacteria and their hosts and its significance. Chapters 4 and 5 provide new insights and exclusive features about the biological relationships of *Wolbachia* with their nematode host. Another fascinating field is *Wolbachia's* role in insects; Chapters 6–7 describe *Wolbachia's* biological significance in insects and insights through their genomic analysis. These two chapters bring additional knowledge, and lessons learned from arthropod *Wolbachia* may shed light on diverse symbiotic associations (parasitism or mutualism) observed in two different organisms. Chapter 8 describes the importance of *Wolbachia* in veterinary filariasis and defines the multiple roles of *Wolbachia* in the pathogenesis, diagnosis and treatment of animal filarial infections, which goes in parallel with studies of *Wolbachia* in human filariasis. Chapter 9 discusses the role of *Wolbachia* in the induction of filarial pathogenesis and critical role of the Toll-like receptor pathways in the host's immune response to these endobacteria.

We hope that the users of this book will enjoy reading the chapters as much as we did!

Ramakrishna U. Rao, St. Louis
Achim Hoerauf, Bonn

Hoerauf A, Rao RU (eds): Wolbachia.
Issues Infect Dis. Basel, Karger, 2007, vol 5, pp 1–14

The Discovery of *Wolbachia* in Arthropods and Nematodes – A Historical Perspective

Wieslaw J. Kozek[a], *Ramakrishna U. Rao*[b]

[a]Medical Sciences Campus, University of Puerto Rico San Juan, P.R.,
[b]Washington University School of Medicine, St. Louis, Mo., USA

Abstract

Collaborative studies between Marshall Hertig, an entomologist, and Samuel Wolbach, a pathologist, on the presence and identification of microorganisms in arthropods, resulted in the discovery of *Wolbachia* in *Culex pipientis* in 1924, although the complete description of *Wolbachia pipientis* was not published until 1936. It has been subsequently demonstrated that *Wolbachia* is widespread in arthropods, infecting about 25–70% of species of insects, and is now known to be a remarkable genetic manipulator of the infected arthropod hosts. Application of electron microscopy to elucidate the structure of nematodes revealed that many filariae (17 species reported to date, including most of the species pathogenic to humans) harbored transovarially transmitted bacterial endosymbionts, subsequently determined as belonging to the *Wolbachia*, clades C, D, and F. The *Wolbachia* are apparently mutualistic endosymbionts required for survival of their hosts and embryogenesis of microfilariae, are present in all larval stages during the life cycle of filarial, and contribute to some of the inflammatory responses and pathological manifestations of filarial infections in the vertebrate hosts. Susceptibility of *Wolbachia* of filariae to certain antibiotics offers an attractive possibility of treatment and control of filarial infections in humans and animals. Recently sequenced genomes of *W. pipientis* (Sanger Institute, UK, and The Institute for Genomic Research, USA) and *Wolbachia* from *Brugia malayi* (New England Biolabs, USA) have opened a new chapter in the studies on *Wolbachia*. The detailed comparisons and the ongoing *Wolbachia* genome sequencing studies in other filarial nematodes and insects could provide the means to fully characterize the structure, composition and the nature of these organisms that play a significant role in mutualism and parasitism.

During the recent years, we have commemorated two important anniversaries related to *Wolbachia*. In 2004, when the 3rd International *Wolbachia* Conference was held on Heron Island, Australia, we observed the 50th anniversary of the

Fig. 1. Samuel Burt Wolbach (1880–1954) in his office, date unknown. (Reproduced from *The Journal of Pathology* by permission of the Pathological Society of Great Britain and Ireland.)

demise of Samuel Burt Wolbach, after whom the bacterium, the subject of our interest and investigations, is named. Now, in 2006, we are celebrating the 70th anniversary of the establishment of the genus *Wolbachia*, and the description of *Wolbachia pipientis* by Marshall Hertig [1]. Collaboration between these two scientists resulted in the discovery of *Wolbachia* and established the groundwork for our current and future studies.

Both scientists lived in an era noted by discoveries of the role that arthropods played in transmission of diseases which have plagued mankind. Patrick Manson's demonstration that *Culex pipiens* was the intermediate host of *Wuchereria bancrofti*, studies on malaria conducted by Ross and Laveran, Chagas' investigations on the transmission of *Trypanosoma cruzi* by the triatomid *Panstrongylus megistus* and work of Bruce, Nabarro, Kline, and Kinghorn and Yorke on the transmission of *Trypanosoma gambinese* and *Trypanosoma rhodesiense* by *Glossina* spp. undoubtedly identified this as an exciting area of research where much could be accomplished and attracted many followers, including Hertig and Wolbach.

Hertig (1893–1978) was an entomologist, trained at the University of Minnesota, who was interested in microbial pathogens of arthropods and in arthropod-transmitted microorganisms. Wolbach (1880–1954; fig. 1) was a Harvard-trained pathologist who had an established reputation as an authority in the area of arthropod-borne infections principally as a result of his studies of Rocky Mountain Spotted Fever and endemic typhus. By 1916, Wolbach has unequivocally implicated *Dermacentroxenus rickettsi* (presently *Dermacentor andersoni*) as the causative agent of Rocky Mountain spotted fever. He studied the tick as the biological transmitter of the disease, described the characteristics of the etiologic *Rickettsia*, and demonstrated its unique ability to parasitize and distend the nuclei in tick tissues. His failure to grow the rickettsiae in cell-free media led him to speculate as to the relationship between the cells of the host and the intracellular parasites. He thus anticipated, as early as the mid-1920s, the concept of the obligatory use by the intracellular organism of the enzyme mechanisms of their hosts. The findings of the League of the Red Cross Societies' Commission, which he headed, on the cause of the typhus fever and demonstration how it was being transmitted was considered as the definite work on the etiology, pathology, and clinical aspects of typhus [2, 3].

Undoubtedly attracted by Wolbach's interest and expertise in the field of arthropod-borne bacterial infections, Hertig came to Harvard to collaborate with Wolbach on the investigation of 'rickettsia-like' organisms present in other insects. Their joint study of the materials collected by Hertig in Peiping, and with Wolbach in the vicinity of Boston, was published in 1924. One of the organisms identified in this study was an unnamed rickettsia-like organism in the gonads of the *C. pipiens* mosquito [4]. However, it was not until 1936 that Hertig, now on the faculty of the Harvard School of Medicine and the School of Public Health, formally established the genus *Wolbachia* in honor of his collaborator and provided a detailed description of the *W. pipientis* organisms (figs 2–4).

Both Hertig and Wolbach would probably be very pleasantly surprised to learn that the unknown rickettsia-like organism, which they first isolated from the common brown mosquito caught in the vicinity of Boston, would be shown to represent a large group of bacteria that infects from 16 to 76% of insects [5, 6] and a large number of other arthropods and invertebrates. They would be impressed by *Wolbachia* ability to ensure its own survival by becoming a genetic manipulator par excellence of the infected hosts, inducing cytoplasmic incompatibility [7], parthenogenesis [8], feminization of male progeny [9] or male killing [10] in the infected hosts. They would probably be equally amazed to learn that some strains of *Wolbachia* have also established their niche in certain parasitic nematodes, some of which are of considerable medical and veterinary importance.

The availability of good electron microscopes and the development and application of refined methodologies to prepare and process biological materials for ultrastructural examination led to the discovery of bacterial endosymbionts, including *Wolbachia*, in nematodes. Most likely the first report of a bacterial entity (probably a *Cardinium* spp.) detected by electron microscopy in tissues of

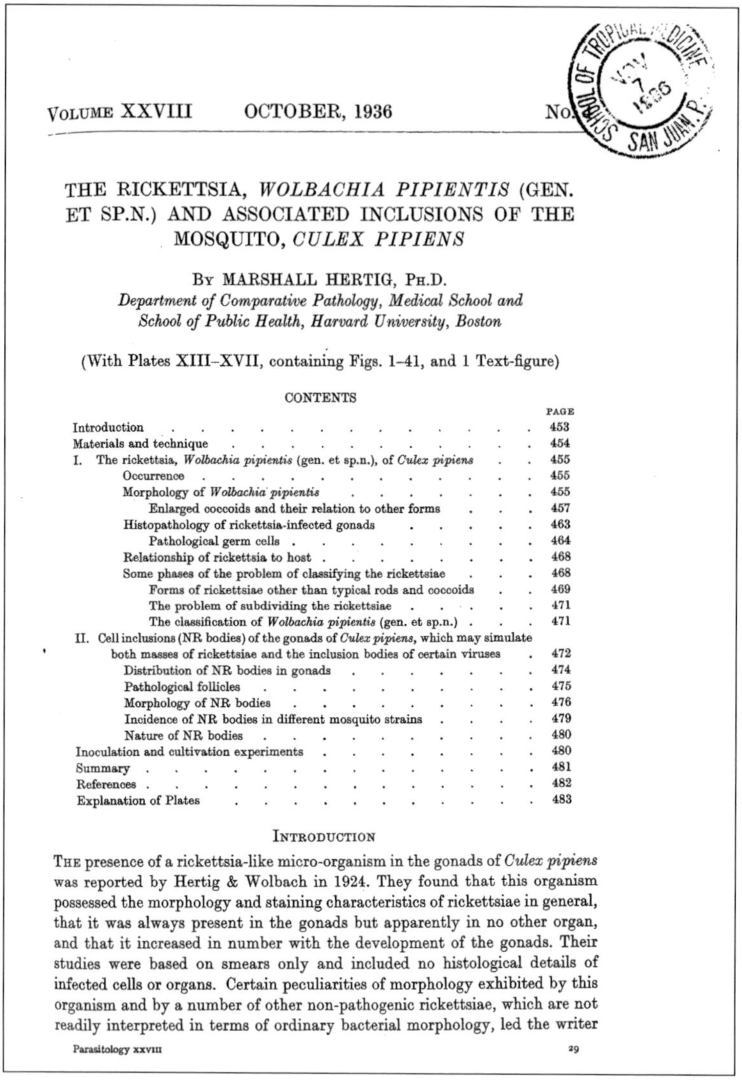

VOLUME XXVIII OCTOBER, 1936 No.

THE RICKETTSIA, *WOLBACHIA PIPIENTIS* (GEN. ET SP.N.) AND ASSOCIATED INCLUSIONS OF THE MOSQUITO, *CULEX PIPIENS*

BY MARSHALL HERTIG, PH.D.
Department of Comparative Pathology, Medical School and School of Public Health, Harvard University, Boston

(With Plates XIII–XVII, containing Figs. 1–41, and 1 Text-figure)

CONTENTS

INTRODUCTION

THE presence of a rickettsia-like micro-organism in the gonads of *Culex pipiens* was reported by Hertig & Wolbach in 1924. They found that this organism possessed the morphology and staining characteristics of rickettsiae in general, that it was always present in the gonads but apparently in no other organ, and that it increased in number with the development of the gonads. Their studies were based on smears only and included no histological details of infected cells or organs. Certain peculiarities of morphology exhibited by this organism and by a number of other non-pathogenic rickettsiae, which are not readily interpreted in terms of ordinary bacterial morphology, led the writer

Parasitology XXVIII 29

Fig. 2. Title page of Hertig's article describing *Wolbachia* [1]. (Reprinted with permission of Cambridge University Press.)

a plant parasitic nematode, *Heterodera* spp., was reported by Shepherd et al. [11]. However, several investigators who conducted ultrastructural studies on filarial nematodes in the late 1960s and early 1970s reported unusual structures in the oocytes or in the hypodermis of the specimens examined, but failed to recognize them as bacterial entities until this identification was suggested by

Wolbachia (gen.n.)

Generic characters. Micro-organisms having the general characteristics of the rickettsiae, and exhibiting not only minute, bacterium-like forms appearing with dark-field illumination as luminous rods and points, but also enlarged forms within the body of which are contained one to several smaller individuals. Type and only known species, *Wolbachia pipientis.*

Wolbachia pipientis (sp.n.)

Specific characters. Wolbachia exhibiting enlarged forms, usually 0·5–1·3 μ in diameter, containing individuals which appear as rings, curved or bent rods, or compact aggregations composed of various numbers and proportions of rod and coccoid forms.

Occurrence. In cytoplasm of germ cells of *Culex pipiens*; distribution apparently that of host. *Type locality,* Boston and vicinity.

Synonym. The rickettsia (unnamed) of *Culex pipiens,* Hertig & Wolbach (1924).

Fig. 3. Hertig's original description of the genus *Wolbachia* and of *W. pipientis* [1]. (Reprinted with permission of Cambridge University Press.)

Vincent and collaborators and confirmed by McLaren and collaborators [reviewed in 12]. Identification of the endosymbionts as belonging to the *Wolbachia* group was not made until 1995 by Sironi et al. [13].

The first detailed studies on the morphology, distribution within the worms and in all stages of the filarial life cycle, mode of reproduction and mode of transmission of bacterial endosymbionts in filariae were made in the mid 1970s on *Onchocerca volvulus*, *Brugia malayi* [14, 15], *Dirofilaria immitis* and *Litomosoides sigmodontis*. These studies have demonstrated that:

1 Some filariae harbored intracellular, pleomorphic bacteria.
2 The bacteria were organotropic, being confined to the hypodermal tissues in all stages of the life cycle of both sexes (figs 5, 6). In the female worms, they were also found in the rachis, oocytes of the female worms, and in the hypodermal precursor cells of the embryos developing in utero (figs 7–10).
3 The bacteria were transovarially transmitted within the life cycle of the filariae.
4 They had two possible modes of reproduction: by binary fission and by a *Chlamydia*-type developmental cycle.
5 No evident morphological pathological effects, in the lateral chords or in the developing filarial embryos, were associated with the presence of the bacteria.

Since numerous experimental studies on the mutualistic association of bacteria in insects and arthropods have already been published, demonstrating that

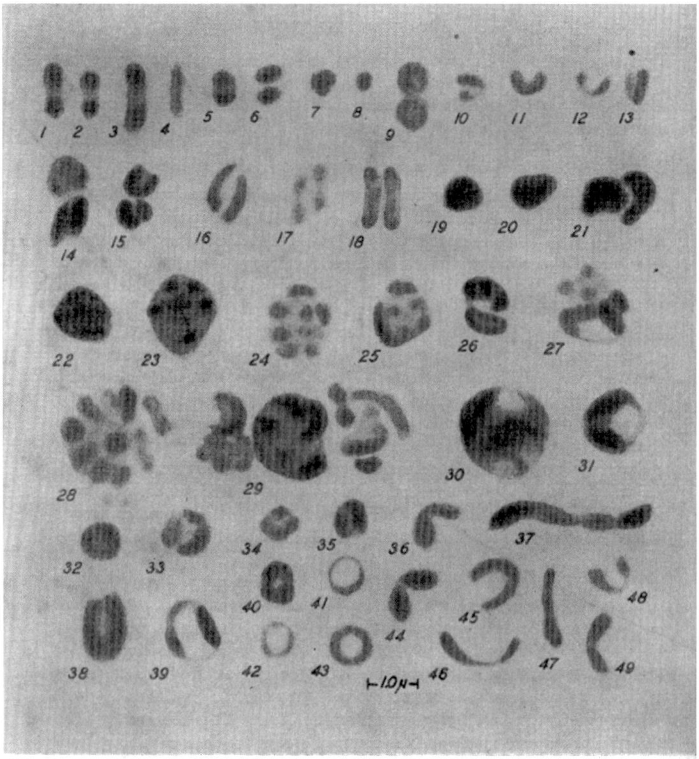

Text-fig. A. Various forms of the rickettsia of *Culex pipiens* from Giemsa-stained smears and sections. Drawn at 10,000 diameters. From ovarian smears, except Nos. 38 and 39 which are from a section of a larva with rickettsiae lying extracellularly in the body cavity. Nos. 1–29, for the most part, drawn from field shown in photomicrograph, Fig. 5.

1–8. "Typical rickettsiae", i.e. small rods and coccoids occurring singly and in pairs.

9, 19–21, 32, 41–43. Enlarged coccoids, intermediate in size, some densely stained throughout (19–21), others with lighter central portions (9, 32) or appearing as rings (41–43).

10, 14, 15, 33–35, 40. Intermediate coccoids, the original outline of which is more or less intact, showing "daughter" forms, of which sharply curved or V-shaped forms and paired rods are common examples.

11–13, 36, 44–49. Curved or V-shaped forms interpreted as having been derived from enlarged coccoids. With the disappearance of the "parental" matrix they tend to straighten out, undergoing elongation and division.

16–18. Paired rods, a very common "daughter" group. They may form ovals (16) or imperfect rings, or may be closely apposed. They may elongate without complete separation (16, 18) or divide to form a tetrad (17).

22–31, 38, 39. Large coccoids with daughter groups in all stages from those exhibiting indistinct punctiform, linear or irregular condensations (22, 23, 25, 30, 31) to those in which rods and coccoids are more or less distinct though still cohering in a compact mass.

9, 14, 15, 21. Enlarged coccoids occurring in pairs, interpreted not as division forms of large coccoids but as the result of enlargement of pairs of small coccoids prior to formation of daughter groups.

37. Filament with several constrictions.

Fig. 4. Hertig's drawings of *W. pipientis* [1]. (Reprinted with permission of Cambridge University Press.)

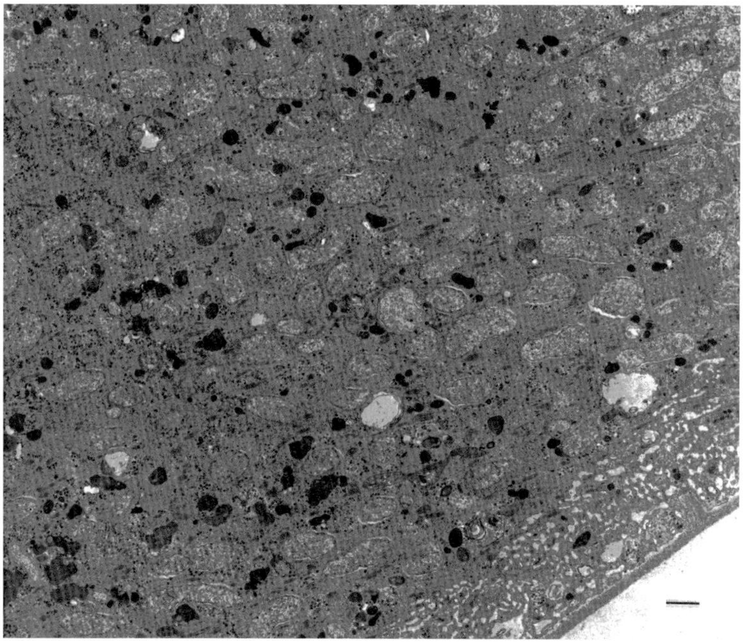

Fig. 5. Numerous *Wolbachia* in the cytoplasm of the lateral chord of *O. volvulus*. Bar = 1 μm.

the induction of aposymbiotic state in the host by elimination of bacterial endosymbionts by surgery, treatment with antibiotics, especially tetracyclines, heat and specific dietary deficiencies produced many untoward effects in the invertebrate hosts, e.g., death, sterility, stunted growth, incomplete formation of the exoskeleton and change in the integrity and coloration of the exoskeleton [16], it appeared obvious that a parallel situation might exist between the bacteria of filariae and their filarial hosts, which led to the following postulates [14, 15]:

1 Lack of obvious pathological effects in the worms suggested a mutualistic relationship between the bacteria and the filariae.
2 This mutualistic association could be exploited for indirect attack on the worms: by killing the worm by chemotherapeutic elimination of the mutualistic bacteria. This would be a novel approach for treatment of filarial infections, using already existing chemotherapeutic agents, which could offer possible applications for the control and/or eradication of filarial infections. If sterility of the worms could be induced by chemotherapy, it would be especially useful in the control of onchocerciasis, where much of the pathology is associated with the host response to microfilariae.

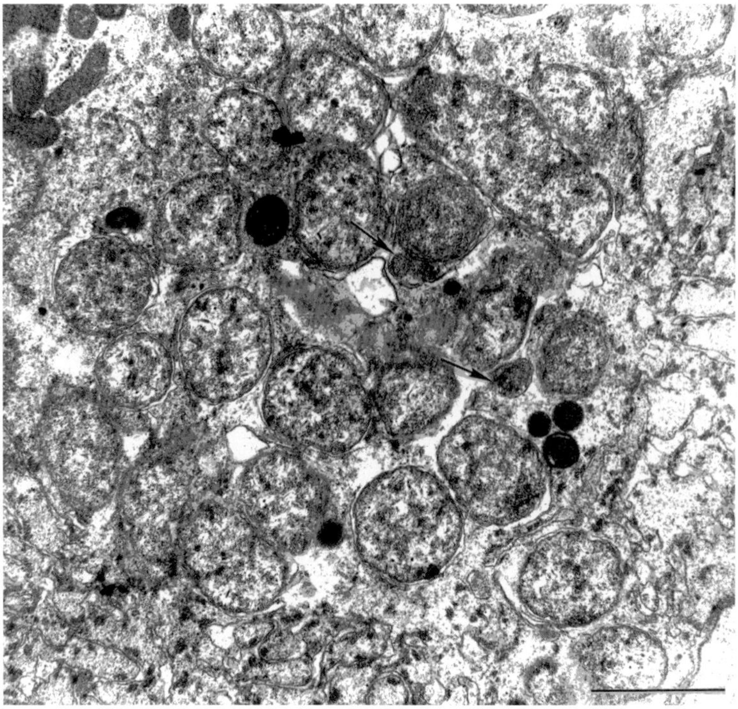

Fig. 6. Pleomorphic *Wolbachia* in the lateral chord of *L. sigmodontis*. Arrows indicate minute bacterial forms. Bar = 1 μm.

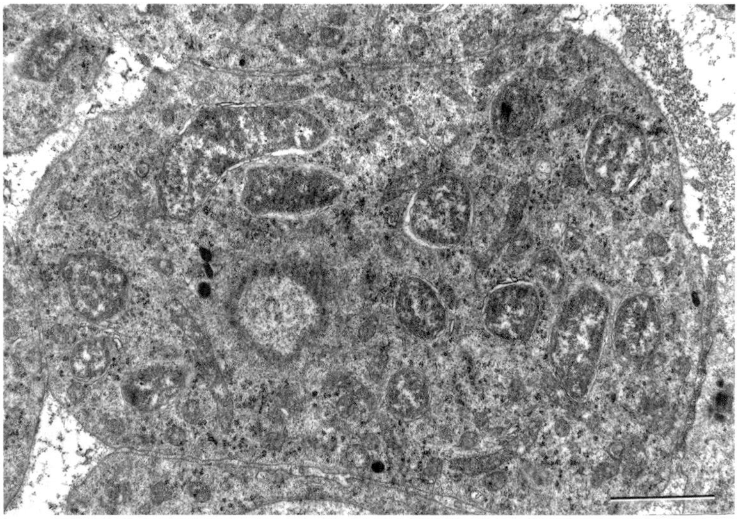

Fig. 7. Numerous *Wolbachia* in the oocyte of *D. immitis*. Bar = 1 μm.

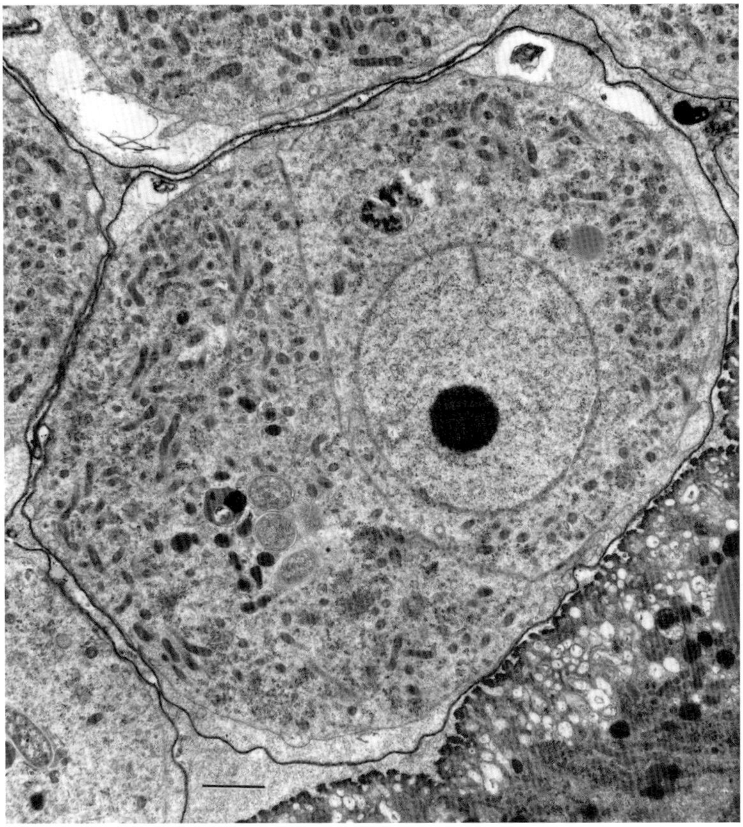

Fig. 8. Accumulation of *Wolbachia* in hypodermal precursor cells of *O. volvulus*; 4- or 8-cell cleavage stage. Bar = 1 μm.

3 The bacteria could contribute to some of the pathological effects of filarial infections observed in the vertebrate hosts.
4 The bacteria and their soluble products should be taken into consideration as inherent contaminants present in all filarial extracts prepared for immunological, biochemical or physiological studies (we can now add to this list molecular biology and genetics).

For the next 17 years, research on bacteria of filariae was limited to very few of laboratories. Our untold story, similar to that of McCall et al. [17], consisted of effort to identify the filarial bacteria by immunological and histochemical techniques available at that time. During the first author's association with the Tulane University International Cooperation in Infectious Diseases Research

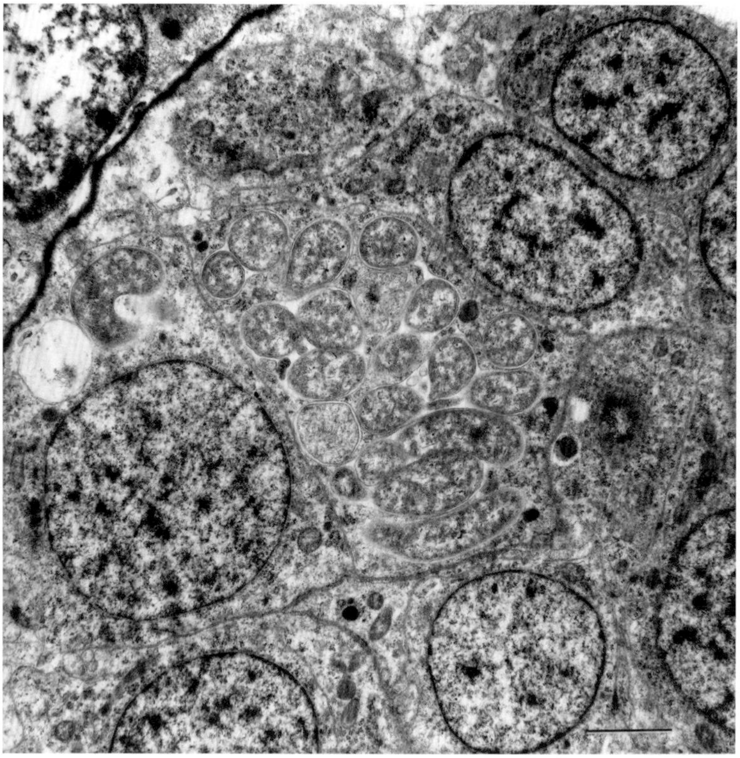

Fig. 9. *Wolbachia* in the hypodermal precursor cells of *O. volvulus*; morula stage. Bar = 1 μm.

program in Cali, Colombia (1977–1984), numerous attempts were made to develop a nematode cell line from *Onchocerca gutturosa*, in which we hoped to culture and maintain the bacteria of filariae [Kozek, unpubl. data]. Subsequently, in Puerto Rico we characterized their histochemical staining characteristics and compared their antigenic relationship to other intracellular bacteria. Indirect fluorescent test did not detect common antigens between the bacteria of *D. immitis* and *Rickettsia* spp. (*R. akari, R. prowazeki, R. tsutsugamushi, R. typhi, R. canada, R. conori*), *Rochelimaea quintana*, *Coxiella burnetti*, *Chlamydia* spp. (*C. trachomatis* and *C. psittaci*), but a weak antigenic relationship to *Ehrlichia* spp. (*E. canis, E. sennetsu, E. risticii* and *E. equi*) was suggested [18, 19], and the complete 1,495 bp 16S rRNA gene of *D. immitis* bacteria was sequenced [19, 20]. With the advent of molecular biology techniques, the Italian group was able to resolve the taxonomic status of the endobacteria of *D. immitis* by identifying their position in the *Wolbachia* group [13].

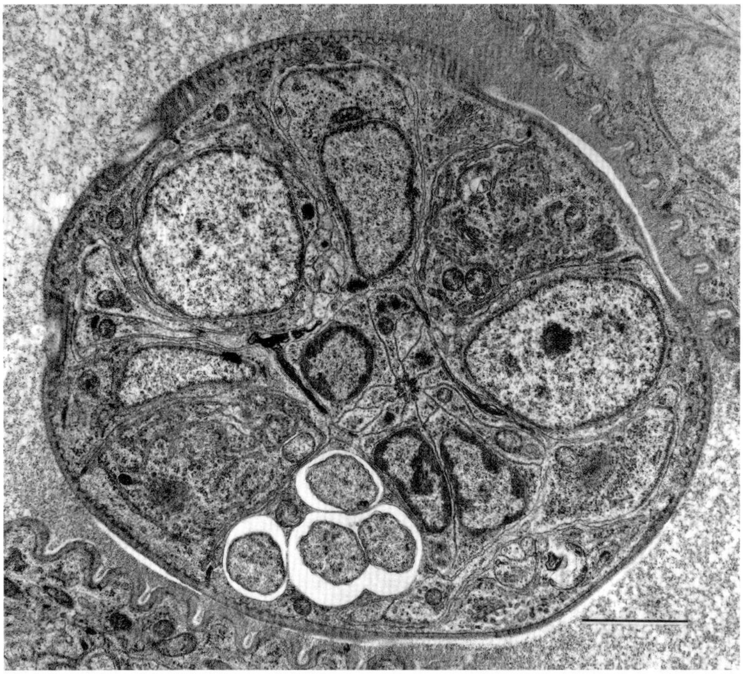

Fig. 10. *Wolbachia* in the hypodermal cells of an in utero microfilaria of *O. volvulus*. Bar = 1 μm.

Current interest in *Wolbachia* of filariae was greatly enhanced by the formation in 1994, under the aegis of the World Health Organization, of a consortium to map the filarial genome. It was somewhat amusing to read the report of the Microbial Genomes III meeting that was held in Chantilly, Va., in February of 1999 [21], reporting that many of the participating investigators found bacterial sequences contaminating their filarial isolates, and referred to previous observations of '…dense bodies' – possibly bacteria…' in filariae of several species. Had a more careful search and analysis of published literature been conducted, these investigators would have found detailed description of these bacteria, even noting their morphological similarity to *Wolbachia*, their distribution within the stages in the life cycle of filariae, vertical mode of transmission, suggestion of their long-term co-evolution with the filarial hosts in a mutualistic association, and indicating that the elimination of these bacteria could be an indirect method to kill the filarial hosts, which had been published more than 25 years previously [14, 15].

The investigations on the *Wolbachia* of filariae have blossomed since the establishment of the filarial genome consortium. The reader is referred to the reviews of Taylor and Hoerauf [22], Rao [23], and Taylor et al. [24], and other chapters of this volume, for the description of the advances that have been made in elucidating many aspects of the distribution of *Wolbachia* among the filariae, its phylogeny and classification, molecular biology, pathogenesis, drug susceptibility and antibiotic treatment of filarial *Wolbachia* as a therapeutic and a control measures. Furthermore, recent sequencing of the genomes of *W. pipientis* [25] and of *Wolbachia* from *B. malayi* [26] will provide a wealth of information and additional areas for research that will help us to understand more fully the nature of *Wolbachia*-filaria relationship. High throughput microarray analysis to study the *Wolbachia* gene expression profiles in *B. malayi* worms clearly showed a significant upregulation of *Wolbachia* genes that are necessary for their survival and protein synthesis [27]. However, the presence of *Wolbachia* in filariae should not be considered as a panacea for the solution of filarial infections, nor as Pandora's box responsible for all the pathological manifestation produced in filarial infections. The *Wolbachia are* inherent in some filaria that infect man and animals, they *do* contribute to *some* pathological manifestations of filarial infections, and they *do* offer a novel approach to chemotherapy and control of filarial infections, especially onchocerciasis. The challenges before us are to identify the specific pathogenic components which *Wolbachia* produce, identify faster-acting antibiotics against *Wolbachia*, mining the genome of *Wolbachia* to identify those biochemical pathways necessary for the survival of their host and their own growth, and apply the results of all these efforts to alleviate the suffering from filarial infections and to hasten the elimination of filariasis from human and animal populations.

Acknowledgements

The authors would like to express their thanks to Ms. Elise Ramsey, Curatorial Assistant of the Warren Anatomical Museum, Francis A. Countway Library of Medicine, Harvard Medical School, Boston, Mass., and Ms. Judy H. Kesenich, Gorgas Memorial Library, Walter Reed Army Institute of Research/Naval Medical Research Center Library, Washington, D.C., for their able assistance in providing many biographical details about Drs. Wolbach and Hertig, respectively. The authors gratefully acknowledge the permission of William Dawson & Sons, Ltd., and Cambridge University Press, to reproduce portions of Dr. Hertig's publication, and the permission of The Pathological Society of Great Britain and Ireland to reproduce the photograph of Dr. Wolbach. Supported, in part, by University of Puerto Rico (Medical Sciences Campus) RCMI Award RR-03051 from the NCRR, NIH, Bethesda, Md., USA.

References

1 Hertig M: The rickettsia, *Wolbachia pipientis* (Gen. et Sp. Nov.) and associated inclusions of the mosquito, *Culex pipiens*. Parasitology 1936;28:453–486.

2 Warren S: Simeon Burt Wolbach, 3rd July 1880–19th March. J Pathol Bacteriol 1954;68:656–657.

3 Farren S, Maddock CL: S. Burt Wolbach, M.D. 1880–1954. Arch Pathol 1955;59:624–630.

4 Hertig M, Wolbach SB: Studies on rickettsia-like micro-organisms in insects. J Med Res 1924;44:329–374.

5 Werren JH, Windsor DM, Guo L: Distribution of *Wolbachia* among neotropical arthropods: Proc R Soc Lond B Biol Sci 2000;267:1277–1285.

6 Jayaprakash A, Hoy MA: Long PCR improves *Wolbachia* DNA amplification: *wsp* sequences found in 76% of sixty three arthropod species. Insect Mol Biol 2000;9:393–405.

7 Stouthamer R, Breeuwer JAJ, Hurst GDD: *Wolbachia pipientis*: microbial manipulator of arthropod reproduction. Ann Rev Microbiol 1999;53:71–102.

8 Stouthamer R, Breeuwer JA, Luck RF, Werren JA: Molecular identification of microorganisms associated with parthenogenesis. Nature 1993;361:66–68.

9 Roussset F, Bouchon D, Pitureau B, Juchault P, Solignac M: *Wolbachia* endosymbionts responsible for various alterations of sexuality in arthropods. Proc Biol Sci 1992;250:91–98.

10 Warren JH, Hurst GD, Zhang W, Breeuwer JA, Stouthamer R: Rickettsial relative associated with male killing in the ladybird beetle *(Adalia bipunctata)*. J Bacteriol 1994;176:388–394.

11 Shepherd AM, Clark SA, Kempton A: An intracellular micro-organism associated with tissues of *Heterodera* sp. Nematology 1973;19:31–34.

12 Kozek WJ: What is new in the *Wolbachia/Dirofilaria* interaction? Vet Parasitol 2005;133:127–132.

13 Sironi M, Bandi C, Sacchi L, Di Sacco B, Damiani G, Genchi C: A close relative of the arthropod endosymbiont *Wolbachia* in a filarial worm. Mol Biochem Parasitol 1995;74:223–227.

14 Kozek WJ, Figueroa Marroquin H: Intracytoplasmic bacteria in *Onchocerca volvulus*. Am J Trop Med Hyg 1977;26:663–678.

15 Kozek WJ: Transovarially-transmitted intracellular microorganisms in adult and larval stages of *Brugia malayi*. J Parasitol 1977;63:992–1000.

16 Buchner P: Endosymbiosis of Animals with Plant Microorganisms. New York, Interscience Publishers, 1965, p 909.

17 McCall JW, Jun JJ, Bandi C: *Wolbachia* and the antifilarial properties of tetracycline. An untold story. Ital J Zool 1999;66:7–10.

18 Cruz-Gonzalez CA: Antigenic characterization of intracellular microorganisms of *Dirofilaria amities*; MS thesis, Puerto Rico, 1989.

19 Kozek WJ, Amigo LA, Cruz-Gonzalez J: Some aspects of the ultrastructural, phylogenetic, immunological and histochemical characteristics of *Wolbachia* of *Dirofilaria immitis*; in Paul AJ (ed): Heartworm Symposium. San Antonio, 2001, pp 215–224.

20 Amigo LA: Molecular characterization and other aspects of 'endosymbionts' of *Dirofilaria immitis*; doctoral dissertation, Puerto Rico, 1999.

21 Pennisi E: DNA Sequences provide grist for microbiologists. Science 1999;283:1105–1106.

22 Taylor MJ, Hoerauf A: *Wolbachia* bacteria of filarial nematodes. Parasitol Today 1999;15: 437–442.

23 Rao RU: *Wolbachia* in worms: endosymbiont of parasitic nematodes. Recent Res Dev Exp Med 2004;1:95–113.

24 Taylor MJ, Bandi C, Hoerauf A: *Wolbachia* bacterial endosymbionts of filarial nematodes. Adv Parasitol 2005;60:246–284.

25 Wu M, Sun LV, Vamethevan J, Reigler M, Deboy R, Brownlie JC, McGraw EA, Martin W, Esser C, Ahmadinejad N, Wiegland C, Madupu R, Beanan MJ, Brinkac LM, Daugherty SC, Durkin SC, Kolonay JF, Nelson WC, Mohamoud Y, Lee P, Berry K, Young MB, Utterback T, Weidman J, Nierman WC, Paulsen IT, Nelson KE, Tettelin H, O'Neill SL: Phylogenomics of the reproductive parasite *Wolbachia pipientis wMel:* a streamlined genome overrun by mobile genetic elements. PLoS Biol 2004;2:327–341.

26 Foster J, Ganatra M, Kamal I, Ware J, Makarova K, Ivanova N, Bhattacharyya A, Kapatral V, Kumar S, Posfai J, Vincze T, Ingram J, Moran L, Lapidus L,Omelchenko M, Kyrpides N, Ghedin

E, Wang S, Goltsman E, Joukov V, Ostrovskaya O, Tsukerman K, Mazur M, Comb D, Koonin E, Slatko B: The *Wolbachia* genome of *Brugia malayi:* endosymbiont evolution within a human pathogenic nematode. PLoS Biol 2005;3:e121.
27 Rao RU, Crosby SD, Mitreva M, Weil GJ: Early effects of doxycycline on *Wolbachia* and parasite gene expression in adult female *Brugia malayi* (abstract 167). Am Soc Trop Med Hyg, 55th Annu Meet, Atlanta, 2006, p 204.

Wieslaw J. Kozek, PhD
Department of Microbiology and Medical Zoology, Medical Sciences Campus
University of Puerto Rico
PO Box 365067
San Juan, PR 00936-5067 (USA)
Tel. +1 787 758 2525, ext. 1351, Fax +1 787 758 4808, E-Mail wkozek@rcm.upr.edu

Hoerauf A, Rao RU (eds): Wolbachia.
Issues Infect Dis. Basel, Karger, 2007, vol 5, pp 15–30

Wolbachia: Evolutionary Significance in Nematodes

Maurizio Casiraghi[a], *Emanuele Ferri*[b], *Claudio Bandi*[b]

[a]Dipartimento di Biotecnologie e Bioscienze, Università degli Studi di Milano Bicocca, [b]Dipartimento di Patologia Animale, Igiene e Sanità Pubblica Veterinaria, Sezione di Patologia Generale e Parassitologia, Università degli Studi di Milano, Milano, Italia

Abstract

Our knowledge of the symbiosis between *Wolbachia* and filarial nematodes has grown rapidly in recent years. Phylogenetic analyses, which highlight a coevolutionary pattern for filarial nematodes and their wolbachiae, and molecular evolutionary analyses, showing no evidence for positive selection in a surface protein, are both concordant with the idea that *Wolbachia* has evolved a mutualistic association with its hosts. There are, however, several open questions regarding the biology and evolution of this symbiotic association. In particular, the actual distribution of *Wolbachia* in filariae and the overall molecular diversity of these bacteria have not yet been completely uncovered. *Wolbachia* is apparently at fixation in positive species of filariae. Other filariae, which are in some cases phylogenetically related to positive species, lack this symbiosis. This picture is intriguing if we consider that *Wolbachia* is thought to be beneficial in positive filariae. If filariae rely on *Wolbachia* for some key biological functions, why should some species have renounced to its presence? We could perhaps suggest that the association between *Wolbachia* and filariae, while being of some usefulness to the nematode, had not evolved to a state of complete dependence of the host, at least in those nematodes that have renounced to this symbiosis.

The history of *Wolbachia* in filarial nematodes is a tale of discovery and rediscovery. At the beginning of the 70s, Harada et al. [1], with electron microscopy (EM), observed round-shaped bodies in tissues of the dog heartworm *Dirofilaria immitis*. Vincent et al. [2] were the first to suggest that these structures were bacteria. In the subsequent years, other EM-based works reporting the presence of intracellular bacteria in filarial nematodes were published [see for instance 3]. In several of these pioneering works, the possibility that these bacteria were Ricketsiales and that they were implicated in some of the clinical manifestation of filariasis was suggested [3]. However, for many years there were no further studies in this direction, and even if other EM observations were reported during the 1980s, the scientific community surprisingly did not pay enough attention to these bacteria and their possible role in filariasis. It was only in the middle of the 1990s that a molecular phylogenetic study led to the identification of the bacteria present in the dog heartworm, *D. immitis*, as close relatives of arthropod *Wolbachia* [4]. In parallel, research in the context of the Filarial Genome Project (started in 1994) contributed to the rediscovery of these endosymbiotic bacteria. It was the beginning of a new course in the field of filariasis research, with different aspects, from chemotherapy to immunology, from genomics to basic biology, re-examined in the light of the presence of *Wolbachia* [5–8].

In filarial nematodes, *Wolbachia* is apparently omnipresent in positive species (i.e. 100% prevalence of *Wolbachia* infection), while other filariae lack this symbiosis [9]. The overall picture of the distribution of these bacteria in filarial nematodes appears to be consistent with a single acquisition of *Wolbachia* on the lineage leading to the subfamilies Onchocercinae and Dirofilariinae. However, several representatives of these two subfamilies could have lost the symbiosis with *Wolbachia*. This is quite an intriguing observation: it is generally thought that *Wolbachia* plays a role as a mutualist in filarial nematodes, in contrast to the apparently prevalent reproductive parasitic role played in arthropods. However, if filarial nematodes rely on *Wolbachia* for some essential biological features, how is it possible that *Wolbachia* infection was lost along some lineages? And, how strong is the interaction between *Wolbachia* and its hosts? At the moment there is no evidence for the presence of *Wolbachia* in nematodes other than the filariae [10], but further screenings are required to understand the real distribution of *Wolbachia* among filarial and other nematodes. A clearer picture of the distribution of *Wolbachia* in nematodes will help answer some of the questions on the origin (and loss) of its association with filariae.

Phylogeny of *Wolbachia* in Filarial Nematodes

Despite the existence of a single valid *Wolbachia* species, i.e. *W. pipientis* [11], vast molecular diversity is observed among the representatives of this genus, which is thus not mirrored in their taxonomic status. Consequently, in the absence of formal descriptions of *Wolbachia* species, the main branches of *Wolbachia* phylogenies have been named 'supergroups' [12]. The first two supergroups identified in insects were named A and B [13]. Two other supergroups, C and D, have

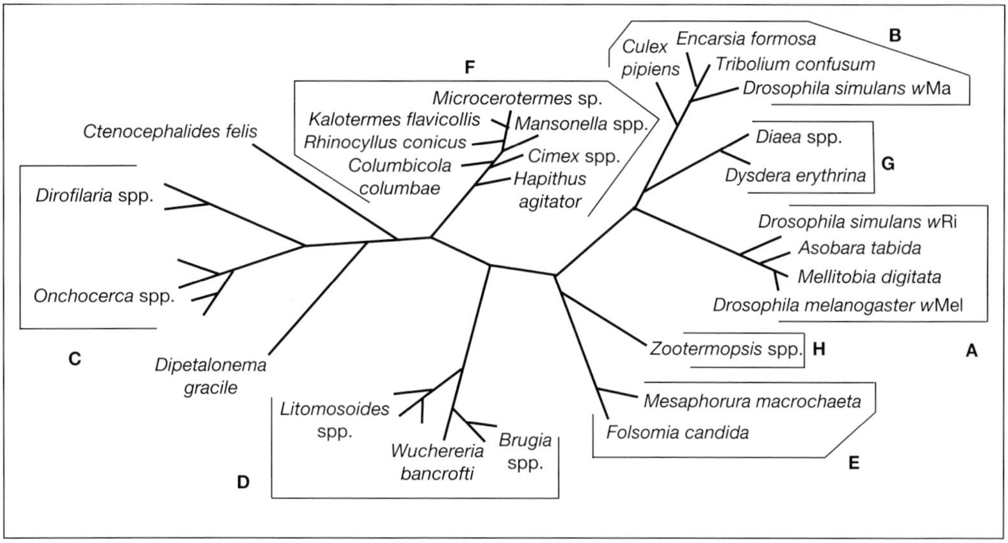

Fig. 1. Unrooted tree of the main *Wolbachia* supergroups. The tree is a representation derived from different works [16, 17, 20] and must be regarded as a possible, but certainly far from definitive, approximation to the 'true tree'. It is also important to note that the branch lengths are not proportional to the evolutionary distance. The *Wolbachia* from the filarial nematode *D. gracile* and the arthropod *Ctenocephalides felis* have not been placed in any of the existing supergroups.

been added following the identification of *Wolbachia* in filarial nematodes [14]. Based on the rates of molecular evolution estimated for bacteria, the evolutionary separation between these four main lineages of *Wolbachia* may have occurred 50–100 million years ago [13, 14]. A further supergroup, E, was then found in arthropods, in particular insect of the order Collembola [15]. At that time, the genus *Wolbachia* was thus divided into supergroups encompassing symbionts of arthropods (A, B and E) or nematodes (C and D). This picture changed dramatically when a further supergroup was described (F), which encompasses *Wolbachia* from both arthropods (in particular termites) and filarial nematodes (species of the genus *Mansonella*) [12]. More recently, new arthropod representatives have been shown to harbor wolbachiae from supergroup F: heteropterans, lice, hippoboscid flies and crickets [16–20]. In addition, the existence of new supergroups has been proposed, G (encompassing *Wolbachia* from spiders) and H (*Wolbachia* from termites, but different from those in supergroup F) [20, 21]. Moreover, further *Wolbachia* molecular diversity has been uncovered in fleas [22–24] and in the filarial nematode *Dipetalonema gracile* [9], even though a supergroup status has not yet been proposed for these new lineages (fig. 1).

The discovery of the diversity encompassed by the genus *Wolbachia* raises the issue of the taxonomic rank that should be attributed to the eight (or even more) supergroups. In the genes coding for the small subunit ribosomal RNA (16S rDNA), the level of divergence between the main *Wolbachia* supergroups is around 3%. It has been proposed that bacterial isolates showing over 3% nucleotide differences in their full length 16S rDNA should be attributed to different species, even if doubts have been advanced on this cut-off approach. Overall, the level of divergence among *Wolbachia* supergroups suggests that these lineages could be elevated at the species rank. For a comparison, we can consider that, based on 16S rDNA sequences, the divergence between the *Wolbachia* of the mosquito *Culex pipiens* (supergroup B) and that of the filarial nematode *Wuchereria bancrofti* (supergroup C) is greater than the difference between rickettsiae assigned to different species such as *Rickettsia rickettsii* and *Rickettsia conorii* or *Rickettsia prowazekii* and *R. rickettsii*.

All of the phylogenetic analyses thus far published have failed to establish the root of the overall *Wolbachia* tree. The branching order of *Wolbachia* supergroups is thus not resolved [12, 24]. In the absence of a robust placement of the root of *Wolbachia* phylogeny, and in view of the growing molecular diversity of the genus *Wolbachia*, decisions on the taxonomic rank of supergroups should be postponed. Among the different supergroups, F deserves particular attention: it is the only supergroup reported to infect both nematodes and arthropods and could thus provide insights into the potential capacity of *Wolbachia* to switch hosts between animal phyla. At the beginning of the evolutionary radiation of *Wolbachia*, transfer of this endosymbiont from arthropods to nematodes (or vice versa) must have occurred. The strict phylogenetic relationship of *Wolbachia* from arthropods and nematodes in the F supergroup shows that a similar transfer might also have occurred more recently and independently from the ancestral host switch. It is thus urgent to acquire information in support of the hypothesis that *Wolbachia* of the F supergroup had a relatively recent host transfer between animal phyla, and in which direction this transfer occurred, i.e. from arthropods to nematodes, or vice versa (fig. 2). Phylogenetic analyses on gene sequences will help to elucidate this issue. Comparison of the genome size and gene order/genome organization will provide further insights into the direction of *Wolbachia* evolution and movement across animal phyla. For example, based on current information, the size of the genome of *Wolbachia* from filarial nematodes (C and D) is smaller than that of *Wolbachia* from insects (A and B). What is the size of the genomes of F wolbachiae? Is the overall organization of these genomes more similar to those from arthropod or nematode wolbachiae?

In summary, the evolutionary history of *Wolbachia* in filarial nematodes appears more complex than the scenarios discussed a few years ago [e.g. 14, 25]. Firstly, several losses of *Wolbachia* appear to have occurred during the evolution

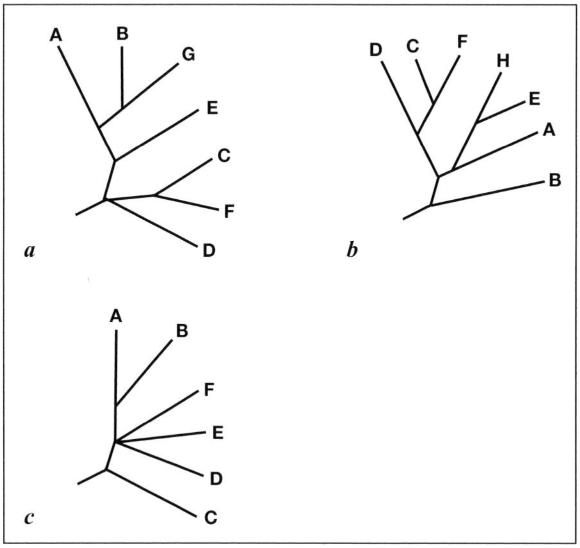

Fig. 2. Rooting of *Wolbachia* tree proposed in some recent works. ***a*** Rowley et al. [20] placed filarial-related wolbachiae plus the mixed group (C, D and F supergroups) as deepest branches (even if the relationships among these branches were not resolved) and arthropod-related wolbachiae (A, B, E and G supergroups) as derived. Among arthropod-related wolbachiae, supergroup E was found as the deepest branch, while B and G the more derived branches. ***b*** Bordenstein and Rosengaus [21] placed arthropod wolbachiae of supergroup B as the deepest branch of the reconstruction, while no choice was made on the other arthropod-related wolbachiae (A, E and H supergroups) and filarial-related wolbachiae plus the mixed group (C, D and F supergroups), which formed anyway two separated clusters, with A and D supergroup as the deepest branches, respectively. ***c*** Fenn and Blaxter [44] placed the filarial wolbachiae of supergroup C as the deepest branch of the reconstruction, while the 'classical' arthropod-related wolbachiae (A and B supergroups) as the more derived. The authors also proposed a cluster with the wolbachiae from supergroups D, E and F, with no clear relationships among them.

of filarial nematodes. Secondly, the coexistence of arthropod and nematode wolbachiae in the F supergroup seems to suggest relatively recent transfer of these endosymbionts between the two types of hosts. The possibility that *Wolbachia* has been acquired more than once by filarial nematodes should thus be considered. On the other hand, the directions of the transfer of *Wolbachia* between animal phyla (i.e. from arthropods to nematodes or vice versa) is still an open question. Another open question regards the loss of *Wolbachia* during the evolution of filariae: why so many losses for a bacterium thought to be beneficial to filarial nematodes? Our current visions and models on the evolution of parasitism and symbiosis are themselves evolving. The relationships among different organisms are the results of a variable balance between costs and benefits, and these results are too complex to be classified into a limited number of categories. The uncovering of the natural history of the reproductive parasite *Wolbachia* has already changed our view of symbiotic relationships; the reconstruction of the overall evolutionary history of parasitic and mutualistic wolbachiae found in arthropods and nematodes will likely introduce a further change in our understanding of symbiotic associations.

Wolbachia Distribution in Filarial Nematodes

Based on the data thus far generated, the presence of *Wolbachia* in filarial nematodes appears to be restricted to the subfamilies Onchocercinae and Dirofilariinae, belonging to the family Onchocercidae (superfamily Filarioidea). The phylogeny of the superfamily Filarioidea is not firmly established. Basic information can be summarized as follows: (1) there is some evidence for a deep branching of the two subfamilies of the Filariidae [9, 26]; (2) within the Onchocercidae (which encompasses eight subfamilies), there is evidence for a deep branching of the subfamilies Setarinae and Waltonellinae, while the evolutionary radiation of the subfamilies Onchocercinae and Dirofilariinae might have occurred after the splits of these two lineages [9]. A recent study showed that the subfamilies Onchocercinae and Dirofilariinae do not form monophyletic lineages: the genera in these subfamilies are frequently intermixed, and might be collected into a single subfamily [9]. We emphasize that all of the filariae causing filariases of medical and veterinary importance are members of the Onchocercinae and Dirofilariinae [26].

The screening for *Wolbachia* in filarial and nonfilarial nematodes has not been extensive, but some filarial groups, and all the nonfilarial nematodes analyzed thus far, have consistently been found devoid of this bacterium [9, 10]. In filarial nematodes, the screening for *Wolbachia* has been carried out in representatives of one out of the two subfamilies of the Filariidae and of four out of the eight subfamilies of the Onchocercidae [9]. The screenings outside filarial nematodes have been even more scattered [10]. In summary, the picture of the distribution of *Wolbachia* within the subfamilies Onchocercinae and Dirofilariinae (family Onchocercidae) shows that there are both positive and negative species, while outside these two subfamilies there are no signs for the presence of *Wolbachia*. Based on these observations, there are two possible alternative scenarios: (1) a single acquisition of *Wolbachia* on the lineage leading to the Onchocercinae/Dirofilariinae with secondary losses that led to the current negative species; (2) the symbioses with *Wolbachia* were established several times in the Onchocercinae/Dirofilariinae subfamilies; in this case, negative species in these subfamilies could be the consequence of either a primitive absence of *Wolbachia* or a secondary loss. Even if available information does not allow to choose between these hypotheses, the close phylogenetic relationship among positive and negative species within the Onchocercinae and Dirofilariinae lineages is coherent with events of *Wolbachia* losses during evolution [9].

On the other hand, the absence of clear evidence supporting the monophyly of *Wolbachia* from filarial nematodes (see above) does not allow to conclude that the symbiosis was established only once (i.e. single acquisition). We could also hypothesize a single acquisition of *Wolbachia* by filarial nematodes, followed by some events of horizontal transmission to arthropods. Again, our attention should be focused on supergroup F, the only supergroup encompassing *Wolbachia* from both filarial nematodes and arthropods, a group clearly supported by the analysis of all of the genes thus far examined [24]. Supergroup F suggests that at least a second event of horizontal transmissions of *Wolbachia* between nematodes and arthropods might have occurred in addition to the original transmission event that presumably established the association in nematodes (or, vice versa, in arthropods). It should be noted that evidence for the possibility of an experimental transfer of *Wolbachia* from *Litomosoides sigmodontis* (naturally harbouring *Wolbachia*) to *Acanthocheilonema viteae* (naturally *Wolbachia*-free) has been reported [27]. After microinjection with wolbachiae purified from *L. sigmodontis*, *A. viteae* worms were successfully implanted into a mammalian host and still found PCR-positive for *Wolbachia* after 8 weeks of maintenance in the mammalian host. These results suggest that horizontal transfers of *Wolbachia* might be possible among filarial nematodes, as it is in arthropods [see for instance 28].

In conclusion, the information on the distribution of *Wolbachia* in filarial nematodes should be interpreted with caution. The extent of *Wolbachia* diversity in filarial nematodes is still unknown, and it is possible that our current conclusions on the distribution of *Wolbachia* in filarial nematodes are deeply biased by insufficient sampling. Indeed, we need to consider that the diversity of filarial nematodes is still not known: birds and reptiles host numerous filarial species for which we have little or no biological and taxonomic information. Is the filaria-*Wolbachia* a mammalian-related matter, or are there other players that have not yet been considered?

The Symbiosis between *Wolbachia* and Its Hosts

Wolbachia is vertically transmitted, from the mother to the offspring. In arthropods, there is evidence for events of horizontal transmission of this bacterium [13, 29], while no such evidence is available for nematodes. The differences in transmission between arthropod and nematode *Wolbachia* has attracted a great deal of interest. There is general agreement that the mode of transmission of a symbiont is a key factor in the evolution of virulence and, more generally, in the kind of relationship that will develop between host and symbiont. Simply put, it is expected that selection will favor increased virulence where parasite transmission is horizontal, where we hypothesize competition among different parasite strains infecting the same host, and where parasite virulence is associated with parasite fecundity and with the chances of transmission. In contrast, a reduction in virulence is expected where symbionts are strictly vertically transmitted from parents to offspring, a situation in which the reproductive success of host and parasite are tightly linked [30]. In such a scenario, selection may also promote a positive contribution of the inherited agent to the host, promoting the development of mutualistic interactions.

As stated previously, the picture described above is a simplification of the way in which selection acts upon vertically transmitted symbionts. The core of the problem is the fact that these symbionts, being inherited, have interests in common with the host individuals that transmit them. *Wolbachia* are maternally inherited and have interests in common with infected female hosts [30]. It is expected that selection should favor beneficial effects towards the host sex responsible for symbiont transmission (i.e. females), whilst beneficial effects on males may follow indirectly. In addition to increasing the fitness of infected females, which are responsible for symbiont transmission, selection can lead to the spread of phenotypes that are detrimental to those hosts which are not involved in symbiont transmission (males or uninfected females). These relationships have been thought in terms of genetic conflicts that are likely to play a central role in the evolution of hereditary symbiosis. In evolutionary terms, there is no intrinsic conflict between being beneficial to females and being disadvantageous to males, and there is no reason not to assume that some hereditary symbionts are at the same time beneficial symbionts as well as reproductive parasites [30].

Which kind of relationships are established between *Wolbachia* and filarial nematodes? The *Wolbachia*-arthropod association is generally less stable than the *Wolbachia*-filarial nematode association. Main differences regard: (1) the rate of horizontal transmission; (2) the prevalence of infection with multiple strains, and (3) the efficiency of vertical transmission. Horizontal transmission and multiple-strain infections appear to be limited to (or at least prevalent in) arthropod wolbachiae, while the efficiency of vertical transmission is apparently higher in filariae. In these nematodes, the distribution of *Wolbachia* (100% prevalence in the infected species and no evidence for multiple infections) and the congruence between the phylogeny of hosts and symbionts are consistent with an obligatory symbiosis. Multiple infections with different strains of a parasite are thought to be a barrier to the evolution of obligatory symbioses, while strict vertical transmission of a symbiont could lead to virulence reduction. These observations thus led to the hypothesis that the association between *Wolbachia* and filarial nematodes is obligatory and not parasitic. Antibiotic treatment experiments on filarial nematodes, analysis of gene sequences and genomic studies provide support for this hypothesis: (1) the antibiotic treatment of filarial nematodes harboring *Wolbachia* is detrimental to the host (see also the following paragraphs); (2) in arthropod wolbachiae there is evidence for positive selection in the *Wolbachia* surface protein (WSP), which suggests an arms race with the host, while positive selection in this protein is not observed in nematode wolbachiae [31, 32]; (3) the genome of the *Wolbachia* from the filarial nematode *Brugia malayi* [7] shows intriguing similarities to the genomes from obligatory, beneficial symbionts of insects, such as *Buchnera aphidicola* in aphids, *Blochmannia floridanus* in carpenter ants and *Wigglesworthia glossinidia* in tsetse flies, even if with some interesting differences.

The Biology of *Wolbachia* in Filarial Nematodes

Localization and Population Dynamics of Wolbachia *in Filarial Nematodes*

Wolbachia has been found throughout all stages of the life cycle of filarial nematodes, although it occurs in varying proportions among individual worms and in different developmental stages [3, 33]. In adult filarial nematodes, *Wolbachia* is mainly found throughout the hypodermal cells of the lateral cords. It has been suggested that the hypodermal lateral cord may be considered a bacteriocyte-like organ, similar to those harboring the endosymbionts *Buchnera* in aphids [34]. In adult females, *Wolbachia* has also been found in the ovaries, oocytes and developing embryonic stages within the uteri, whereas they have not been observed in the male reproductive system.

Even if *Wolbachia* has been found in all developmental stages of filarial nematodes, the level of bacterial infection changes during filaria development [33, 34]. In *B. malayi*, the ratio *Wolbachia* DNA/filarial DNA is constant in the mosquitoes from the stage of microfilaria to the stage L_3, but shows a dramatic increase within 7 days after the infection of the mammalian host. The ratio of symbiont DNA/host DNA further increases during development throughout L_4. Following the final molt, the increase in *Wolbachia* content in males is limited, while females appear to be the site of a very intense multiplication of *Wolbachia*. This hyperabundance of *Wolbachia* in the adult females suggests some association with oogenesis/embryogenesis, which agrees with the evidence from antibiotic treatment experiments leading to inhibition of embryogenesis and microfilarial production (see below).

A rapid and considerable increase in the *Wolbachia* load is thus observed in L_3s during the 1st week following infection of the mammalian host. It is also interesting to note that, according to a recent study [33], single-round PCR detected *Wolbachia* in only about 30% of the L_3s examined. It is curious that a similar percentage of inoculated filarial larvae had been reported in other studies to establish infection in the gerbil. This raises the question of the real efficiency of *Wolbachia* transmission in filariae from the mother to the offspring: is it possible that transmission efficiency is different in different oocytes, or indeed is it less than 100%? Can this differential transmission to be related to the success of vertebrate infection by L_3? If this is true, we could perhaps suggest that the role of *Wolbachia* in microfilariae and during the phase in the invertebrate hosts is different than that in the vertebrate hosts. It is also possible that *Wolbachia* plays its role as a mutualist only during development/survival/reproduction in the vertebrate hosts. There is presently no information which could suggest a possible role of *Wolbachia* in microfilariae, or during L_1–L_3 development.

It is difficult to understand how the increase in *Wolbachia* number is regulated: is it under the control of the bacteria, or the worm, or both? During nematode growth, the lateral cord cells likely expand, generating new space that bacteria can occupy. Available space can be a limiting factor for bacteria division [34]. However, the rapid increase in *Wolbachia* number in the first 7 days after infection cannot be explained in terms of available space only. Studies on the dynamics of *Wolbachia* within the host are ongoing and will help to clarify these still open issues. Interestingly, individual worms show a great variability in their bacterial load, which may reflect a dynamic change of bacterial population size over time [35]. This variability deserves further attention, since it could be linked to a selective advantage in terms of longevity or fecundity in worms carrying more bacteria. Indeed, in a recent study comparing different isolates of *Onchocerca volvulus* in West Africa the bacterial load has been associated with the virulence level of the filaria. A significantly greater ratio of *Wolbachia* DNA to nuclear DNA has been found in the 'severe' compared to the 'mild' isolates of the parasite. Since the 'severe' isolates are responsible for severe ocular disease leading to river blindness, these results confirm the role of *Wolbachia* in the pathogenesis of ocular onchocerciasis [36].

Antibiotic Treatments on Wolbachia *Hosts*

The effects of antibiotic treatment on filarial nematodes suggest a close, and possibly obligatory, mutualistic relationship between *Wolbachia* and its hosts. Studies have provided interesting clues for the possible role of *Wolbachia* in the biology of filarial nematodes: *Wolbachia* appears to be linked to filarial nematode development, reproduction, molt and possibly, to the long-term survival of the worm. However, the results of experiments with tetracycline (and other antibiotics) must be interpreted carefully. It is indeed possible that the *Wolbachia*-filaria relationship is a mutualistic system (and that tetracycline interferes with this system), but it is also possible that tetracycline has a direct effect on the nematode.

There is general agreement that antibiotic therapy has antifilarial effects due to its activity against *Wolbachia*, because antibiotics have no effect on the *Wolbachia*-negative filaria *A. viteae* and because there is evidence that the antibacterial effects precede the antifilarial effects. In a recent study, it has been shown that irradiation of *B. malayi*, leading to reduction in *Wolbachia* loads, has effects on worm motility, viability, development and embryogenesis similar to antibiotic treatment, suggesting a possible role of the endosymbionts on these features [37]. On the other hand, it has been shown that a chemically modified tetracycline interferes with the L_3–L_4 molt in *B. malayi* apparently without depletion of *Wolbachia* [38]. It must be emphasized, however, that the apparent lack of effects on *Wolbachia* was tested only by nonquantitative PCR.

The consequences of antibiotic treatment appear to be the result of worm dependence on the *Wolbachia* infecting the hypodermal lateral cord cells, since the effects are, in general, seen in both male and female filarial nematodes [33], supporting the hypothesis of a bacteriocyte-like role for these wolbachiae in filarial nematodes [34]. This is apparently contradicted by a study reporting a different effect of tetracycline in male and female worms [39]. However, since tetracycline appears to interfere with filarial molt, it is possible that the difference in the effect on males and females was related to the difference in their molting times.

The recent publication of the genome from *Wolbachia* of the filaria *B. malayi* [7] has provided several clues on the possible role of *Wolbachia* in filarial biology. According to Foster et al. [7], *Wolbachia* likely provides its hosts with heme (the prosthetic group of cytochromes, catalase and peroxidase), a molecule that could be critical in filarial development and reproduction. Indeed, it is possible that nematode molt and reproduction are regulated by ecdysteroid-like hormones. On the other hand, there is no evidence for genes for heme biosynthesis in the *B. malayi* genome. Thus, *Wolbachia*-derived heme could play a key role in filarial biology. The effect of antibiotic treatment on filarial nematodes could therefore be due to heme depletion influencing filarial viability, molting, development, and microfilarial production [7].

Evidence from Wolbachia *Molecular Evolution and Genomics*

The wide range of phenotypes caused by *Wolbachia* on arthropod hosts (from feminization of genetic males to cytoplasmic incompatibility) led to studies aimed at reconstructing the phylogeny of these bacteria. The main purpose of these initial investigations was to provide insights into the evolution of these phenotypes, but a by-product of these studies has been the accumulation of a number of gene sequences. Initial phylogenies, inferred from 16S rDNA sequences, showed the monophyly of the genus *Wolbachia*, but did not allow to resolve the branching order of the main lineages of this bacterium. Successive analyses, performed with cell-cycle gene *ftsZ*, and the surface protein gene *wsp*, revealed that phylogenies inferred from different genes were not always congruent (this was the first evidence for genetic recombination in arthropod wolbachiae), and that the same phenotypes were caused by phylogenetically unrelated wolbachiae. In the case of arthropod *Wolbachia*, detailed analyses have led to the conclusion that genetic recombination indeed occurred among *Wolbachia* lineages [40, 41]. However, the mechanisms at the base of recombination are still a matter of speculation.

An analysis of the rate of *Wolbachia* recombination (including arthropod and nematode *Wolbachia*) was published by Jiggins [42]. The different methods used by Jiggins consistently showed a high rate of recombination in arthropod *Wolbachia*, showing values comparable to the rate of the horizontally transmitted bacteria of the genus *Cowdria*, which suggests that different *Wolbachia* strains had come into contact more frequently than presumed. On the other hand, the rate of recombination in nematode *Wolbachia* was found to be, as expected, roughly null. Further support for the absence of recombination in filarial *Wolbachia* came from the comparison of phylogenies based on different genes, including *dnaA* gene sequences. In order to explain these data, Jiggins [42] proposed two hypothesis: (1) the rate of recombination is low in taxa with reduced rates of horizontal transmission; (2) the rate of recombination of arthropod *Wolbachia* is low, but recombinant genotypes are strongly favored by natural selection [42].

As stated above, arthropod *Wolbachia* generally acts as a reproductive parasite, and this agrees with the evidence for horizontal transfer between host species. Nematode *Wolbachia* appears strictly vertically transmitted and is considered a mutualist. We can thus expect that *Wolbachia* is under different selective pressures in arthropods and nematodes. Two studies used the same approach to detect positive selection in genes from arthropod and nematode *Wolbachia* [31, 32]. In nematode *Wolbachia*, no evidence was detected for positive selection in any of the genes analyzed, including the gene coding for the WSP. On the other hand, positive selection (i.e. an excess of nonsynonymous substitution) was detected in the gene coding for WSP of arthropod *Wolbachia*, as well as in a portion of the *ftsZ* coding gene. Jiggins et al. [31] interpreted the positive selection found in the arthropod *Wolbachia wsp* gene as the effect of an arms race between arthropod host and *Wolbachia* parasites. Baldo et al. [32]

also proposed that positive selection could derive from the need of a molecular adaptation to different cellular environments after horizontal host shifts of arthropod *Wolbachia*. Notably, these two hypotheses are not mutually exclusive [32].

The publication of the complete genome sequence of *Wolbachia* from *B. malayi* (*w*Bm), the second completely sequenced genome of *Wolbachia*, and the first of a nematode *Wolbachia*, allowed a series of evolutionary comparative analyses with genomes of other intracellular bacteria, including the genome of *Wolbachia* from *Drosophila melanogaster* (*w*Mel) [7]. The smaller genome size of *w*Bm (1,080,084 nucleotides of *w*Bm versus 1,267,782 nucleotides in *w*Mel), and the smaller number of predicted coding genes (806 of *w*Bm versus 1,271 of *w*Mel) might reflect the molecular evolution of a mutualistic bacterium compared to a parasitic one: the parasitic *w*Mel could have retained genes required for host infection and manipulation, while these genes have been lost in the mutualistic *w*Bm. In contrast to *w*Mel, *w*Bm does not contain prophages and has a reduced level of repeated DNA, which may reflect a stronger selection for repeat loss in *w*Bm. It has also been suggested that the high level of repetitive DNA in *w*Mel, relative to *w*Bm, may reflect the parasitic lifestyle of these bacteria. However, recent studies showed that bacteriophage WO is more widespread in arthropod *Wolbachia* than previously recognized, occurring in at least 89% (35/39) of the sampled genomes. In the region encoding a putative capsid protein, the recombination rate is higher than that of any known recombining genes of the *Wolbachia* genome. Gene transfer by bacteriophages could drive significant evolutionary change in the genomes of intracellular bacteria that are typically considered highly stable and prone to genomic degradation.

Like *R. prowazekii* and *R. conorii*, *w*Bm and *w*Mel genomes have undergone considerable gene losses in many metabolic pathways and cell envelope biogenesis relative to other α-proteobacteria (apparently both wolbachiae are unable to synthesize lipid A, commonly found in parasitic proteobacteria, including members of the genus *Rickettsia*). These observations are in agreement with the hypothesis that claims that gene acquisition and gene loss could play a major role in promoting the spectrum of interactions between bacteria and their hosts. In *w*Mel, the presence of a considerable amount of repetitive DNA and of an apparently active system of DNA recombination may be responsible for the extensive events of genome shuffling that have eliminated the colinearity between *w*Bm and *w*Mel genomes [7]. From the evolutionary point of view, it is interesting to note that *Wolbachia* genomes apparently do not contain regions which could have been acquired recently from the host, while there is evidence that the genome of the beetle *Callosobruchus chinensis* contains a genome fragment derived from *Wolbachia*. Similarly, a recent work detected a DNA sequence in the genome of *O. volvulus* which is thought to derive from the genome of *Wolbachia* [43]. We might perhaps suggest that the acquisitions of genome fragments from *Wolbachia* by the nematode hosts could explain the symbiont losses that apparently happened several times during filarial evolution.

Genomic analyses also provided support to the idea that *w*Bm is an obligate and possibly beneficial symbiont: *w*Bm may provide its host with essential metabolites like riboflavin and flavin adenine dinucleotide, heme, purine and pyrimidine nucleotides [7]. It is interesting to note that, in contrast to other intracellular symbionts, like *Buchnera*, there is no evidence that suggests a role for *Wolbachia* in providing amino acids to its hosts; on the contrary, it is likely that *w*Bm receives amino acids necessary for bacterial metabolism from *B. malayi* [7].

A final interesting evolutionary observation deriving from the comparative genomics of bacterial endosymbionts is that the different proteobacteria symbionts independently evolved distinctive strategies in the symbiont-host relationship. For example, *B. aphidicola* and *B. floridanus* conserved almost all amino acid biosynthesis pathways, supplying their insect hosts with amino acids. In contrast, *w*Bm, *w*Mel and *W. glossinidia* have lost most of the amino acid biosynthesis pathways, but have retained biosynthesis of nucleotides and some coenzymes [7, 6]. We can conclude that proteobacteria-metazoa relationships evolved in different directions, and obligatory symbionts, even though not phylogenetically distant, developed independent means of interaction with their hosts.

References

1 Harada R, Maeda T, Nakashima A, Sadakata M, Anso M, Yonomine K, Otsuji Y, Sato H: Electronmicroscopical studies on the mechanism of oogenesis and fertilization of *Dirofilaria immitis*; in Sasa M (ed): Recent Advances in Research on Filariasis and Schistosomiasis in Japan. Baltimore, University Park Press, 1970, pp 99–121.

2 Vincent AL, Portaro JK, Ash LR: A comparison of the body wall ultrastructure of *Brugia pahangi* with that of *Brugia malayi*. J Parasitol 1975;61:570–576.

3 Kozec WJ: Transovarially-transmitted intracellular microorganisms in adult and larval stages of *Brugia malayi*. J Parasitol 1977;63:992–1000.

4 Sironi M, Bandi C, Sacchi L, Di Sacco B, Damiani G, Genchi C: A close relative of the arthropod endosymbiont *Wolbachia* in a filarial worm. Mol Biochem Parasitol 1995;74:223–227.

5 Bandi C, Slatko B, O'Neill SL: *Wolbachia* genomes and the many faces of symbiosis. Parasitol Today 1999;15:428–429.

6 Wu M, Sun LV, Vamathevan J, Riegler M, Deboy R, Brownlie JC, McGraw EA, Martin W, Esser C, Ahmadinejad N, Wiegand C, Madupu R, Beanan MJ, Brinkac LM, Daugherty SC, Durkin AS, Kolonay JF, Nelson WC, Mohamoud Y, Lee P, Berry K, Young MB, Utterback T, Weidman J, Nierman WC, Paulsen IT, Nelson KE, Tettelin H, O'Neill SL, Eisen JA: Phylogenomics of the reproductive parasite *Wolbachia pipientis w*Mel: a streamlined genome overrun by mobile genetic elements. PLoS Biol 2004;2:E69.

7 Foster J, Ganatra M, Kamal I, Ware J, Makarova K, Ivanova N, Bhattacharyya A, Kapatral V, Kumar S, Posfai J, Vincze T, Ingram J, Moran L, Lapidus A, Omelchenko M, Kyrpides N, Ghedin E, Wang S, Goltsman E, Joukov V, Ostrovskaya O, Tsukerman K, Mazur M, Comb D, Koonin E, Slatko B: The *Wolbachia* genome of *Brugia malayi*: endosymbiont evolution within a human pathogenic nematode. PLoS Biol 2005;4:E121.

8 Brattig NW, Bazzocchi C, Kirschning CJ, Reiling N, Buttner DW, Ceciliani F, Geisinger F, Hochrein H, Ernst M, Wagner H, Bandi C, Hoerauf A: The major surface protein of *Wolbachia* endosymbionts in filarial nematodes elicits immune responses through TLR2 and TLR4J. Immunol 2004;173:437–445.

9 Casiraghi M, Bain O, Guerrero R, Martin C, Pocacqua V, Gardner SL, Franceschi A, Bandi C: Mapping the presence of *Wolbachia pipientis* on the phylogeny of filarial nematodes: evidence for symbiont loss during evolution. Int J Parasitol 2004;34:191–203.

10 Bordenstein SR, Fitch DHA, Werren JH: Absence of *Wolbachia* in nonfilariid nematodes. J Nematol 2003;35:266–270.

11 La Scola B, Bandi C, Raoult D: Genus *Wolbachia* Hertig 1936; in Brenner DJ, Krieg NR, Staley JT, Garrity GM (eds): Bergeys Manual of Systematic Bacteriology. New York, Springer, 2005, pp 138–143.

12 Lo N, Casiraghi M, Salati E, Bazzocchi C, Bandi C: How many *Wolbachia* supergroups exist? Mol Biol Evol 2002;19:341–346.

13 Werren JH, Zhang W, Guo LR: Evolution and phylogeny of *Wolbachia*: reproductive parasites of arthropods. Proc R Soc Lond B 1995;261:55–63.

14 Bandi C, Anderson TJC, Genchi C, Blaxter ML: Phylogeny of *Wolbachia* in filarial nematodes. Proc R Soc Lond B 1998;265:2407–2413.

15 Vandekerckove TTM, Watteyne S, Willems A, Swing JG, Mertens J, Gillis M: Phylogenetic analysis of the 16S rDNA of the cytoplasmic bacterium *Wolbachia* from the novel host *Folsomia candida* (Hexapoda, Collembola) and its implications for wolbachial taxonomy. FEMS Microbiol Lett 1999;180:279–286.

16 Rasgon JL, Scott TW: Phylogenetic characterization of *Wolbachia* symbionts infecting *Cimex lectularius* L. and *Oeciacus vicarius* Horvath (Hemiptera: Cimicidae). J Med Entomol 2004;41: 1175–1178.

17 Sakamoto JM, Feinstein J, Rasgon JL: *Wolbachia* infections in the Cimicidae: museum specimens as an untapped resource for endosymbiont surveys. Appl Environ Microbiol 2006;72:3161–3167.

18 Covacin C, Barker SC: Supergroup F *Wolbachia* bacteria parasitise lice (Insecta: Phthiraptera). Parasitol Res 2007;100:479–485.

19 Panaram K, Marshall JL: F supergroup *Wolbachia* in bush crickets: what do patterns of sequence variation reveal about this supergroup and horizontal transfer between nematodes and arthropods? Genetica, in press.

20 Rowley SM, Raven RJ, McGraw EA: *Wolbachia pipientis* in Australian spiders. Curr Microbiol 2004;49:208–214.

21 Bordenstein S, Rosengaus RB: Discovery of a novel *Wolbachia* super group in Isoptera. Curr Microbiol 2005;51:393–398.

22 Fischer P, Schmetz C, Bandi C, Bonow I, Mand S, Fischer K, Buttner DW: *Tunga penetrans*: molecular identification of *Wolbachia* endobacteria and their recognition by antibodies against proteins of endobacteria from filarial parasites. Exp Parasitol 2002;102:201–211.

23 Gorham CH, Fang QQ, Durden LA: *Wolbachia* endosymbionts in fleas (Siphonaptera). J Parasitol 2003;89:283–289.

24 Casiraghi M, Bordenstein SR, Baldo L, Lo N, Beninati T, Wernegreen JJ, Werren JH, Bandi C: Phylogeny of *Wolbachia pipientis* based on *gltA*, *groEL* and *ftsZ* gene sequences: clustering of arthropod and nematode symbionts in the F supergroup, and evidence for further diversity in the *Wolbachia* tree. Microbiology 2005;151:4015–4022.

25 Bandi C, Trees AJ, Brattig NW: *Wolbachia* in filarial nematodes: evolutionary aspects and implications for the pathogenesis and treatment of filarial diseases. Vet Parasitol 2001;98:215–238.

26 Anderson RC: Nematode Parasites of Vertebrates – Their Development and Transmission. Wallingford, CAB International, 2000, pp 467–590.

27 Hartmann N, Stuckas H, Lucius R, Bleiss W, Theuring F, Kalinna BH: Trans-species transfer of *Wolbachia*: microinjection of *Wolbachia* from *Litomosoides sigmodontis* into *Acanthocheilonema viteae*. Parasitol 2003;126:503–511.

28 Poinsot D, Bourtzis K, Markakis G, Savakis C, Mercot H: *Wolbachia* transfer from *Drosophila melanogaster* into *D. simulans*: host effect and cytoplasmic incompatibility relationships. Genetics 1998;150:227–237.

29 Kikuchi Y, Fukatsu T: Diversity of *Wolbachia* endosymbionts in heteropteran bugs. Appl Environ Microbiol 2005;69:6082–6090.

30 Bandi C, Dunn AM, Hurst GDD, Rigaud T: Inherited microorganisms, sex-specific virulence and reproductive parasitism. Trends Parasitol 2001;17:88–94.

31 Jiggins FM, Hurst GDD, Yang Z: Host-symbiont conflicts: positive selection on an outer membrane protein of parasite but not mutualistic rickettsiaceae. Mol Biol Evol 2002;19:1341–1349.

32 Baldo L, Bartos JD, Werren JH, Bazzocchi C, Casiraghi M, Panelli S: Different rates of nucleotide substitution in *Wolbachia* endosymbionts of arthropods and nematodes: arms race or host shifts? Parassitologia 2002;44:179–187.

33 McGarry HF, Egerton GL, Taylor MJ: Population dynamics of *Wolbachia* bacterial endosymbionts in *Brugia malayi*. Mol Biochem Parasitol 2004;135:57–67.

34 Fenn K, Blaxter M: Quantification of *Wolbachia* bacteria in *Brugia malayi* through the nematode lifecycle. Mol Biochem Parasitol 2004;137:361–364.

35 Taylor MJ, Bandi C, Hoerauf A: *Wolbachia* bacterial endosymbionts of filarial nematodes. Adv Parasitol 2005;60:245–284.

36 Higazi TB, Filiano A, Katholi CR, Dadzie Y, Remme JH, Unnasch TR: *Wolbachia* endosymbiont levels in severe and mild strains of *Onchocerca volvulus*. Mol Biochem Parasitol 2005;141: 109–112.

37 Rao R, Moussa H, Vanderwaal RP, Sampson E, Atkinson LJ, Weil GJ: Effects of gamma radiation on *Brugia malayi* infective larvae and their intracellular *Wolbachia* bacteria. Parasitol Res 2005;97:219–227.

38 Rajan TV: Relationship of anti-microbial activity of tetracyclines to their ability to block the L3 to L4 molt of the human filarial parasite *Brugia malayi*. Am J Trop Med Hyg 2004;71:24–28.

39 Casiraghi M, McCall JW, Simoncini L, Kramer LH, Sacchi L, Genchi C, Werren JH, Bandi C: Tetracycline treatment and sex-ratio distortion: a role for *Wolbachia* in the moulting of filarial nematodes? Int J Parasitol 2002;32:1457–1468.

40 Jiggins FM, Schulenburg JHGVD, Hurst GDD, Majeurs MEN: Recombination confounds interpretations of *Wolbachia* evolution. Proc R Soc Lond B 2001;268:1423–1427.

41 Baldo L, Bordenstein S, Wernergreen JJ, Werren JH: Widespread recombination throughout *Wolbachia* genomes. Mol Biol Evol 2006;23:437–449.

42 Jiggins FM: The rate of recombination in *Wolbachia* bacteria. Mol Biol Evol 2002;19:1640–1643.

43 Fenn K, Conlon C, Jones M, Quail MA, Holroyd NE, Parkhill J, Blaxter M: Phylogenetic relationships of the *Wolbachia* of nematodes and arthropods. PLoS Pathog 2006;2:E94.

44 Fenn K, Blaxter M: *Wolbachia* genomes: revealing the biology of parasitism and mutualism. Trends Parasitol 2006;22:60–65.

Claudio Bandi
Dipartimento di Patologia Animale, Igiene e Sanità Pubblica Veterinaria
Sezione di Patologia Generale e Parassitologia
Università degli Studi di Milano
Via Celoria 10, IT–20133 Milan (Italy)
Tel. +39 02 5031 8093, Fax +39 02 5031 8095, E-Mail claudio.bandi@unimi.it

Hoerauf A, Rao RU (eds): Wolbachia.
Issues Infect Dis. Basel, Karger, 2007, vol 5, pp 31–51

Wolbachia Endosymbionts: An Achilles' Heel of Filarial Nematodes

Achim Hoerauf, Kenneth Pfarr

Institute for Medical Microbiology, Immunology and Parasitology, University Clinic Bonn, Bonn, Germany

Abstract

Filarial worm infections of humans cause morbidity and even death in developing countries of the tropics. Current antifilarial drug therapies target only the first-stage larvae, requiring many years of annual/biannual treatment. Another problem with controlling filarial infections is the lack of any alternative drugs that can be used in the current mass drug administration programs should resistance develop. *Wolbachia*, endosymbiotic bacteria that are found in most of the human filarial worms are excellent targets for the discovery of new antifilarial drugs because of their requirement for worm embryogenesis, development and adult survival. Targeting of *Wolbachia* with antirickettsial drugs has lead to the recommendation of doxycycline for use on an individual basis and may be recommended in areas where resistance to current drugs may develop. More evidence that eliminating the endobacteria reduces adverse reactions to current drug therapies and even reduces early stages of pathology is also accruing. Thus, research is underway to discover new drugs, preferably those already approved for use in humans, that have antiwolbachial activity and work in a shorter time and which can be given to all members of the population.

'Two urns on Jove's high throne have ever stood,
The source of evil one, and one of good;
From thence the cup of mortal man he fills,
Blessings to these, to those distributes ills;
To most he mingles both.'

Homer

Filarial nematode infections are endemic in more than 90 countries of the tropics. In these countries, 200 million individuals are infected and 1.3 billion people are at risk of infection [1–3]. The number of people infected is higher than the estimate given here due to problems with identifying infected individuals, insufficient funding and political instability [4, 5]. Human filarial infections can cause two pathologies depending on the nematode. Lymphatic filariasis (LF) is caused by *Wuchereria bancrofti* and *Brugia* spp. in Africa, India and South East Asia. LF can cause recurrent debilitating fevers, lymphangitis and elephantiasis, and is currently estimated to affect 44 million individuals [3, 6]. Lymphangitis and elephantiasis are the result of the host immune response factors that are induced when adult worms die in the lymphatic vessels. Onchocerciasis, or river blindness, is caused by *Onchocerca volvulus* in Africa and parts of Central and South America. In contrast to LF, host inflammatory reactions to the dead *O. volvulus* larvae (microfilariae, MF) in the skin and eye causes skin disease, Sowda (leopard skin) and visual impairment. Past vector control programs have reduced transmission in West Africa with a resulting decrease in the number of patients with onchocerciasis. However, in Central Africa the number of onchocerciasis patients is conservatively estimated at 37 million, and is probably higher [2, 4, 7, 8].

Filarial infections have a negative impact on the infected individuals and the community in developing countries. This includes the health of the individual, the stigmatization of infected individuals and the economic impact due to loss of productivity [9–11]. Thus, filarial infections are a major public health problem in developing countries. It was thought that filarial infections only lead to high morbidity, not mortality. However, several reports have produced more results which correlate an increased mortality rate for persons infected with *O. volvulus* in comparison to uninfected members of the same community [12–15].

Biology of Filarial Infections

Adult nematodes are sexually dimorphic and reside in the lymphatic vessels (LF) or in subcutaneous nodules (onchocerciasis). *W. bancrofti* and *Brugia* spp. worms can survive in the human hosts for 4–6 years [16]. *O. volvulus* worms can survive 14–15 years [17]. During this time the adults mate and the females, over their life span, release millions of MF which circulate in the blood (LF) or migrate through the skin and eyes (onchocerciasis). All filarial nematodes have an obligate insect vector required for development and transmission to humans. LF is transmitted by several genera of mosquitoes (*Aedes*, *Anopheles*, *Culex*, and *Mansonia*). Onchocerciasis is transmitted by black flies of the genus *Simulium*. Current drugs used to control transmission primarily target the MF.

There are indications that ivermectin (IVM), the drug used to control onchocerciasis transmission, may have prophylactic activity against infective stage larvae. Additionally, repeated doses of IVM can also lead to a block in embryo release, probably a result of paralysis of the uterine muscles, and therefore degeneration of embryos in utero [18–21]. There is one recent report that presents evidence for macrofilaricidal activity after 3 years of quarterly IVM treatments [22]. However, the percentage of worms killed by this intense IVM treatment that can be directly attributed to IVM is ca. 13–15%. As will be explained later, modeling of onchocerciasis control has shown that only 3% of a population needs to be infected in order for onchocerciasis to become re-established as a public health problem [23]. The drug diethylcarbamazine citrate (DEC), used to treat LF, does have some macrofilaricidal effects [24–27]. However, both drugs require multiple doses and have a very low efficacy for macrofilaricidal activity.

Past and Current Control Efforts

Efforts to control filarial infections focus on three goals: (1) reduce the intensity of infection to levels such that morbidity is below levels where the disease is a public health problem; (2) regional elimination which leads to the prevention of new infections; and (3) eradication of the worldwide incidence of infection. The WHO has helped affected countries to develop control programs to accomplish the first step. The Onchocerciasis Control Programme in West Africa (OCP) used insecticides to control the black fly vectors in all participant countries. During the OCP, several countries also administered IVM to affected villages. This successful program ended in 2002 and resulted in a reduction in the incidence of blindness due to onchocerciasis by one third [28, 29]. OCP also demonstrated that filarial disease could be controlled and lead to the development of newer programs based on mass drug administration (MDA) of antifilarial drugs [29, 30].

Currently, there are two programs to control onchocerciasis, the African Programme for Onchocerciasis Control (APOC; http://www.who.int/blindness/partnerships/APOC/en/), administered in areas of participating countries hyperendemic for onchocerciasis, and the Onchocerciasis Elimination Programme for the Americas (OEPA; http://www.cartercenter.org) [25, 31]. Efforts to control LF are managed under the Global Programme for the Elimination of Lymphatic Filariasis (GPELF; http://www.filariasis.org) [6, 30, 32].

All programs use annual or semi-annual administration of antimicrofilarial drugs to interrupt transmission. APOC and OEPA administer IVM (provided free by Merck) in regions where onchocerciasis alone is endemic. The GPELF administers IVM and albendazole (ALB, provided free by GlaxoSmithKline) [33] in areas where LF is co-endemic with onchocerciasis, and DEC and ALB where LF is monoendemic. DEC can be inexpensively made in endemic countries. Administration of ALB is expected to improve compliance in the programs by curing patients of gastrointestinal helminth infections which, in contrast to blocking the transmission of filariae, offer a result that can be immediately seen by treated individuals [34–36].

OEPA, administered in Latin America, is expected to be successful as onchocerciasis is hypoendemic in this region. This is mainly due to the fact that the *Simulium* species in Latin American countries are suboptimal at transmitting the infection [29]. The affected populations are also small enough that it is feasible to administer IVM twice a year. Thus, current mathematical models predict that the current levels of coverage, if maintained, will be sufficient to eliminate onchocerciasis from the New World [37]. However, even when transmission is shown to be halted, IVM must continue to be taken for the lifespan of the worms, i.e. 15 years.

Onchocerciasis in the rest of the world may prove harder to eliminate. Outside of the Americas, the *Simulium* flies are the optimal vectors. Semi-annual IVM treatment is not feasible as the affected regions are much larger than those found in the Americas. *O. volvulus* can live up to 15 years, requiring IVM administration for this number of years, even *after* transmission has been interrupted. Mathematical modeling has predicted that a minimum of 25–35 years is required to break transmission in areas hyperendemic for onchocerciasis when coverage is 65% of the infected population [38]. These models do not yet take into account any resistance to IVM that may develop in the nematodes.

Compounding the problems of controlling onchocerciasis is that some regions are co-endemic for *O. volvulus* and *Loa loa* (transmitted by *Chrysops* spp.) [39, 40]. IVM is also effective against *L. loa*. However, it must be used with caution as patients with high *L. loa* MF loads may develop, due to the rapid killing of the *L. loa* MF, encephalitis that can potentially be fatal [41, 42]. Such severe side effects effectively prohibit the current MDA programs from areas co-endemic for *L. loa*, areas that could be a source for new infections if they border areas where onchocerciasis is eliminated as a public health problem.

There are other indications that IVM and DEC treatment alone will not be sufficient. In onchocerciasis, a study in Cameroon of sites that had received IVM therapy for 10–12 years found infection rates of 2–3% at the end of treatment, or when treatment was interrupted. This level of infection is enough to re-establish the infection within a few years [23]. Similar results have been seen with LF. A follow-up study of DEC treatment of *W. bancrofti* infections on a remote island in French Polynesia showed that despite 34 years of drug administration, about 4% of the population is still infected. This includes children as young as 2 years of age, which are clearly new infections [43].

To date, the exact mode of action of DEC is still unclear, but host factors are clearly needed. These factors appear not to require T cells or the complement system, but do require components of the innate immune system [44, 45]. Recent research has demonstrated the in vivo requirement of inducible nitric oxide synthase and the cyclooxygenase pathway for killing of MF by DEC [46]. Once the biochemical action of DEC is understood, a potential threat of resistance to DEC can be better evaluated. IVM binds to glutamate-gated chloride channels of nematodes. This binding leads to a slow, but permanent opening of the channel which leads to the blocking of the affected muscle tissue [47]. IVM is also a substrate for the multi-drug resistance protein P-glycoprotein [48].

Drug resistance is becoming more of a concern to the MDA programs. The study in French Polynesia speculated that the persistent infection could be due to worms which were resistant to DEC [43]. There have also been reports of 'low responders' to IVM in foci in Ghana, which could be the presages of IVM resistance in onchocerciasis [49, 50]. A recent genetic examination of *W. bancrofti* MF before and after IVM/ALB treatment demonstrated that there is selection for a mutation in the actin gene that is associated with resistance to ALB in veterinary nematodes [51]. A separate study in Latin America was able to detect live *W. bancrofti* in 38% of the patients after repeated doses of DEC [52]. In *O. volvulus*, a loss of polymorphism in several loci after a single IVM treatment has also been shown. One of the proteins for which a loss of polymorphism was seen is P-glycoprotein, a protein that is associated with IVM resistance in veterinary helminths [53, 54]. Clearly, selection is taking place when treating with IVM. The development of resistance to IVM or DEC would be catastrophic to APOC/OEPA and GPELF as these are the only drugs currently approved for MDA.

As will be explained in the following paragraphs, other than moxidectin and doxycycline, no new antifilarial drugs have been developed, and other drugs which have been studied are either less effective than the current ones, or they are too toxic (table 1) [55–58]. Moxidectin has been shown to be macrofilaricidal in animal studies not dealing with *Onchocerca* species [59]. It is known that onchocercae are usually not as susceptible to drug action as other species. Even if moxidectin would show improved activity over IVM, it might be a short-lived replacement for IVM in the case that IVM resistance becomes prevalent, because it targets the same glutamate-gated chloride channel that IVM does [47] and is *also* a substrate for P-glycoprotein [60]. Resistance to moxidectin has already been found in the veterinary nematodes *Haemonchus contortus* and *Ostertagia circumcincta* [61]. To date, clinical trials with moxidectin have proven safe in humans [62]. However, moxidectin has been toxic when given to dogs that have a single nucleotide mutation in their P-glycoprotein gene that leads to an amino acid change [63]. It is not unimaginable that a similar mutation may be present in the human gene.

Because of the current long treatment times required and the specter of resistance to DEC or IVM developing, new drugs that are macrofilaricidal or faster acting are needed. Ideally, these drugs would already be registered for use in humans, and they should not be more expensive to produce than the current

Table 1. Antifilarial drugs past and present [30, 55]

Drug	Activity	Comment/contraindication
ALB, mebendazole citrate (benzimidazoles)	Reduction/interruption of embryogenesis	Has only weak antifilarial activity on its own. Given in combination with DEC and IVM to treat nonfilarial helminth infections. Resistance development a concern [51].
Amocarzine	Microfilaricidal	No longer available.
DEC	Microfilaricidal; potentially macrofilaricidal (40% efficacy) [52]	Used to treat LF. Due to severe adverse effects (Mazzotti reaction) no longer used to treat onchocerciasis.
Doxycycline	Interruption of embryogenesis; macrofilaricidal in LF (88% efficacy) [27, 126]	Complete and permanent inhibition of embryogenesis in onchocerciasis after a 6-week treatment at 100 mg/day. 70–80% macrofilaricidal activity over 1–2 years in LF and onchocerciasis when given for 6 or 8 weeks at 200 mg/day [27, 126].
IVM	Microfilaricidal; partial interruption of embryogenesis after frequent application	Sole drug approved to combat onchocerciasis as part of APOC and OEPA. Used in the GPELF in areas co-endemic for onchocerciasis. Nonresponders in current programs may indicate resistance is developing [49].
Levamisole	Anthelmintic and immunostimulant	Has no direct effect on MF or adult worms. Immunostimulatory activity does not enhance the action of IVM or ALB [56].
Melarsoprol (Mel W)	Potentially macrofilaricidal	Too toxic for MDA.
Metrifonate	Microfilaricidal activity	Currently not available. Efficacy is much lower than that of DEC.
Moxidectin	Potentially macrofilaricidal in animals [59]	In trials for LF and onchocerciasis. Resistance seen in veterinary nematodes [61].
Suramin	Reduction/interruption of embryogenesis; macrofilaricidal	Must be given intravenously, requiring administration in a clinic. High toxicity in a cohort study (18%), probably due to overdosing, exemplifies the narrow therapeutic window.

drugs available. Current research which has lead to a new antifilarial therapy focuses on the intracellular, endosymbiotic bacteria of the filarial worms.

Wolbachia, Targets for a Novel Chemotherapy against Filariasis

Since the 1970s, it has been known that filarial worms contain endosymbiotic bacteria. Morphologically, these endobacteria resemble rickettsial endosymbionts [64]. In the worm, the endobacteria are found in the hypodermis, the

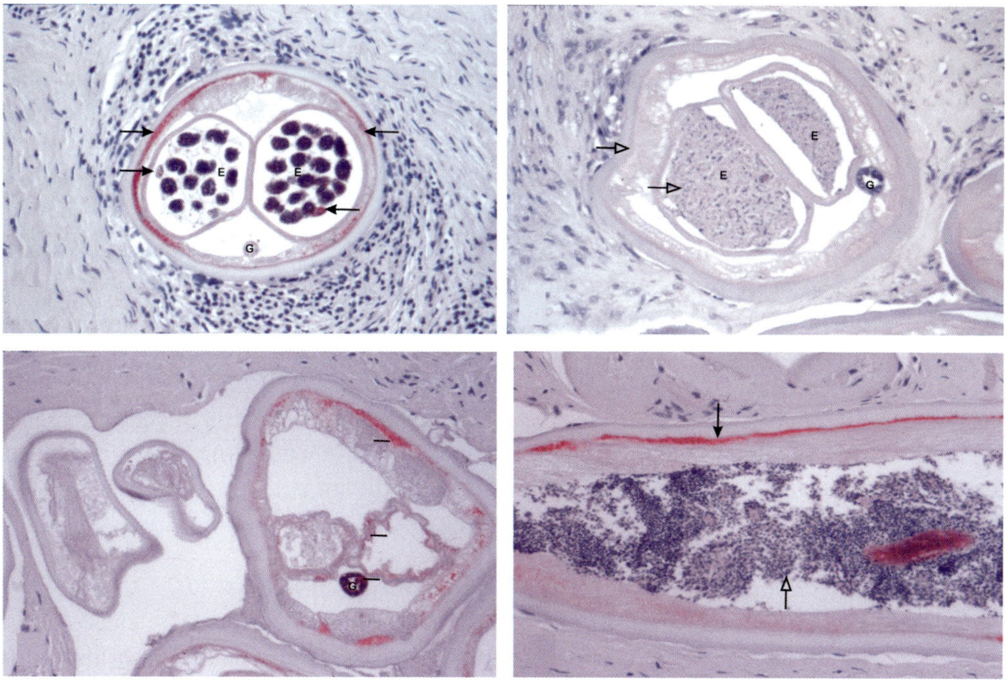

Fig. 1. Immunohistochemistry can be used to monitor the efficacy of antiwolbachial therapy (***a, b, d***) and live versus dead worms (***c***). ***a*** Cross-section of a female worm showing that endobacteria (arrows) are in the hypodermis and embryos (E), not in other tissues. ***b*** Cross-section of a female worm from a patient that was treated for 6 weeks with doxycycline. *Wolbachia* are no longer seen in the hypodermis (open arrows), resulting in degenerated embryos (E). ***c*** *Wolbachia* aspartic protease is a marker for live worms. Strong staining of the worm on the right is seen in the gut (G) and other tissues of the worm (lines). In contrast, no staining is seen in the worm sections on the left, which has degraded tissues within the cuticle. ***d*** Ivermectin treatment does not affect *Wolbachia* (arrow), but neoplasms sometimes develop after multiple rounds of ivermectin treatment (open arrow). Images are the kind gift of Prof. Dr. Dietrich W. Büttner, Bernhard Nocht Institute for Tropical Medicine, Hamburg, Germany.

oocytes, and in all embryonic and larval stages (fig. 1a) [65–69]. Molecular techniques identified these endobacteria as *Wolbachia* and have shown that they are closely related to the same genus found in many insects [70, 71]. The endobacteria are transmitted from the females to the next generation via the ova (vertical transmission). No transmission to other species, as is seen with insect *Wolbachia*, has been described. These two facts suggest a mutualistic symbiotic relationship. Consistent with this, the phylogenies of the *Wolbachia* strains from the various nematodes closely parallel that of the worm host [71–73]. The *Wolbachia* of filarial nematodes appear not to be under any genetic selection pressures, probably due to their mutualistic lifestyle, nor has there been any evidence of recombination found. This contrasts with the positive genetic selection and recombination documented in arthropod *Wolbachia* [74, 75]. As nematode *Wolbachia* lack both of these factors, the rapid development of resistance to any drug found to have antiwolbachial activity would be hindered.

The genome sequence from the *Wolbachia* of *Brugia malayi* has been completed, providing researchers with information about the requirements of the endobacteria [76, 77]. The endobacteria lack the ability to make all but one amino acid, yet retain the ability to make nucleotides, heme, riboflavin and flavin adenine dinucleotide, all of which may be metabolites provided by the endobacteria to the nematode as part of the symbiosis [76]. While *Mansonella perstans* and *L. loa* worms do not have *Wolbachia* [78–80], *Brugia* spp., *Mansonella ozzardi*, *O. volvulus*, and *W. bancrofti* contain these endosymbionts [81, 82], thus opening up a new era of chemotherapeutic drug discovery against the three major causative agents of human filariasis.

Wolbachia Are Vital to the Biology of Filarial Worms

In several different animal filarial infections, both in natural hosts and models for human infections, antiwolbachial treatment with tetracycline has demonstrated that the *Wolbachia* are essential to worm biology [83–85]. In all tests, the reduction in the number of MF in the blood can be traced to a block in embryogenesis which is preceded by the depletion of the endobacteria by tetracycline or other antirickettsial drugs, i.e. rifampicin [86–89]. In studies with *Onchocerca ochengi*, a filarial nematode of cattle, tetracycline treatment leads to death of the adult worms [87, 90]. The block in embryogenesis seen after antiwolbachial treatment is a direct effect of the depletion of the *Wolbachia* as tetracycline treatment of animals infected with *Acanthocheilonema viteae*, a filarial nematode without *Wolbachia*, has no effect on embryogenesis or worm vitality [85]. In *B. malayi*, the endobacteria can also be depleted from the nematodes by irradiation in a dose-dependent manner. This depletion leads to the characteristic phenotype seen in *Wolbachia*-depleted worms, i.e. blocked embryogenesis [91].

Treatment of infective larvae with tetracycline in vitro also leads to an inhibition in their ability to molt [92]. A block in larval molting due to the loss of the endosymbionts would explain the inability of *Brugia pahangi* (a filarial nematode of cats that can infect rodents) larvae to develop into adult worms in Mongolian gerbils that are treated with tetracycline prior to and during the larval molting period [93]. However, the antimolting effect of tetracycline may be a direct toxic effect of tetracycline on the larvae, e.g. by damage to the mitochondria or calcium chelation. This is supported by a recent report that showed that the use of a synthetic tetracycline lacking antimicrobial activity still blocks the larval molt [94].

Wolbachia Are Strong Inducers of the Immune Response

In the past few years, the importance of these endobacteria as sources of pathology of filarial disease has become apparent. In animal models of filariasis, pathology, including lymphedema in rhesus monkeys, correlates with increased levels of circulating *Wolbachia* DNA or proteins. The appearance of these *Wolbachia* products corresponds with the loss of MF. Presumably killing of MF by the host releases *Wolbachia* into the blood [95, 96]. A link between the most severe filarial pathology of onchocerciasis and *Wolbachia* has also been demonstrated with an in vivo model for blindness in mice [97]. The development of blindness in mice after injection of worm extract into the cornea is dependent upon *Wolbachia* as *O. volvulus* extract depleted of *Wolbachia* does not induce blindness. Blindness is also not induced when these mice do not have a functional Toll-like receptor (TLR) 4 molecule [98]. TLRs are key receptors in the innate immune reaction to foreign antigens [99]. Continuing work with this model has shown a requirement for the TLR cofactor myeloid differentiation factor D 88 [100], TLR2, but not TLR9 in the development of ocular pathology [Gillette-Ferguson and Pearlman, pers. commun.].

The current dogma for how pathology develops is that the infected patient has a hyperimmune response to worm antigens. In vitro experiments have shown that extracts of *B. malayi* and *O. volvulus* induce proinflammatory cytokines by macrophages and monocytes [101, 102]. The induction of these cytokines is dependent on *Wolbachia* as extracts from worms depleted of endobacteria by tetracycline or from *A. viteae*, a filaria with no *Wolbachia*, do not lead to significant induction of inflammatory cytokines [97, 102]. The failure of *A. viteae* extract to induce proinflammatory cytokines is not because *A. viteae* is not a source of antigen, but rather due to the absence of *Wolbachia* in these worms. Support for *Wolbachia* being powerful inducers of inflammation has also been demonstrated using Aa23 cells, a *Wolbachia* containing insect cell line [103]. Extracts of this cell line also induce an inflammatory response from macrophages, but Aa23 cells cured of *Wolbachia* do not [102]. Furthermore, neutrophils, innate inflammatory cells, migrate in vitro only to worm extracts containing *Wolbachia*. In nodules extirpated from deer infected with two different *Onchocerca* spp., neutrophils are only found surrounding worms which contain *Wolbachia*, but not in the species that is devoid of the endobacteria, and in onchocercomas from doxycycline-treated patients which harbor *Wolbachia*-depleted *O. volvulus* [104]. A potential molecule that is mediating the inflammatory response is *Wolbachia* surface protein (Wsp), which elicits a strong inflammatory response via TLR2 and TLR4 [105]. Wsp was seen to localize in macrophages that were incubated with MF, demonstrating that the human host immune system does come into contact with *Wolbachia* antigens.

***Wolbachia* Contribute to the Adverse Reactions Seen after Microfilaricidal Treatment**

Some patients develop severe systemic inflammatory responses immediately after receiving antifilarial drugs [20, 106, 107]. Reactions include fever, headache, dizziness, myalgia, arthralgia and enlargement of the lymph nodes. Corresponding to the adverse reactions, increases in serum levels of cytokines and cytokine receptors indicative of systemic inflammation (interleukin-6, -10 and the tumor necrosis factor receptors) have been measured. The severity of these reactions has been associated with the microfilarial level before treatment [41, 42, 108] and the antihelmintic drug used to kill the MF [20]. Such adverse reactions may hinder MDA programs as they reduce participation in the programs, which will reduce the coverage needed to eliminate filarial diseases [109].

Several recent reports have linked the increase in *Wolbachia* antigens (Wsp)/DNA to adverse reactions to antifilarial drug therapy. Serum was taken from Indonesian patients infected with *B. malayi* before treatment with DEC and for several time points after. *Wolbachia* DNA was detected in sera from all 3 patients who experienced severe reactions after DEC and 1 patient who experienced a moderate adverse reaction [110]. An ongoing study of the effect of treating *B. malayi* patients with doxycycline prior to the administration of DEC indicates that patients who received doxycycline had milder adverse reactions than placebo patients. These same doxycycline patients also had statistically lower levels of serum IL-6 [Supali et al., pers. commun.].

A recent study with *W. bancrofti* patients also showed a significant reduction in moderate adverse reactions in patients who received doxycycline 3 weeks prior to receiving IVM/ALB. At the end of doxycycline treatment, the levels of MF were reduced in treated patients. As expected, the MF had fewer *Wolbachia*. The moderate adverse reactions in the placebo doxycycline group correlated with the levels of *Wolbachia* released into the blood after receiving the antifilaricide. Patients who had doxycycline treatment prior to IVM and ALB had significantly lower inflammatory cytokines, which correlated with *Wolbachia* levels [111].

An association of *Wolbachia* with adverse reactions after antifilarial therapy has also been seen in *O. volvulus* infections [112]. Higher levels of *Wolbachia* DNA were detected in the blood of patients who had moderate to severe adverse reactions after receiving DEC or IVM. Patients who received DEC had significantly higher levels of bacterial DNA in the blood. In addition to statistically higher levels of endosymbiont DNA in the blood, patients that received IVM also had significantly higher levels tumor necrosis factor-α. This cytokine correlated with bacterial DNA levels.

Filarial nematodes without *Wolbachia* also cause pathology/adverse reactions. The best example of this is seen with some *L. loa* infections [41, 113, 114]. However, pathology is generally only seen in *L. loa* patients with very high MF levels (>15,000 MF/ml) in the blood or spinal/cerebral fluid. In contrast, pathology may develop in onchocerciasis patients with just 50 MF/skin snip (roughly equivalent to 25 MF/ml). Secondly, *M. ozzardi*, which does contain *Wolbachia* [78, 79, 81], does not cause any pathology in most infected patients [115, 116]. This may reflect the location of the adult worms in the peritoneal cavity rather than the lymphatic vessels, or the presence of MF in the blood rather than in the skin or the eyes. Additionally, *M. ozzardi* MF are smaller than *Brugia* spp., *W. bancrofti*, and *O. volvulus* and therefore may have fewer *Wolbachia*.

Nevertheless, these studies show that in patients infected with *Brugia* spp., *W. bancrofti* or *O. volvulus*, *Wolbachia* are major mediators of adverse reactions seen after antifilarial therapy with DEC or IVM. Depletion of *Wolbachia* with doxycycline prior to antifilarial therapy reduces the severity of the adverse reactions observed in patients infected with these species. Notably, the treatment regime with doxycycline needed to achieve a reduction in adverse reactions and amicrofilaremia is shorter (3 weeks) than that needed to produce a macrofilaricidal effect (6 or 8 weeks) [111, 117, 118; Hoerauf et al., unpubl. results].

Antiwolbachial Treatment: A New Tool against Human Filariasis

Based on the promising results from animal experiments, and given that doxycycline is a registered drug, open phase IIa studies have been carried out since 1999 in the rainforest zone of Central Ghana in villages that are hyperendemic for onchocerciasis. Patients participating in the studies have received 100 or 200 mg/day doxycycline for several weeks. Patients also received IVM after doxycycline therapy as part of the implementation of APOC. The antiwolbachial activity has been monitored by evaluating MF in the skin and nodules, and of adult worms in extirpated nodules 2–24 months after commencement of therapy. Samples were analyzed by microscopy, immunohistology and PCR.

As seen in animal studies, after a 6-week course of doxycycline treatment (100 mg/day) the endobacteria were eliminated from the worms (fig. 1a–b) [69, 88]. Again, loss of the endobacteria resulted in a block in embryogenesis. This block in embryogenesis has been documented 24 months after commencement of therapy [69], making doxycycline the first antifilarial agent which completely blocks embryogenesis without serious side effects. MF levels in the skin reflect the block in embryogenesis, with 90% of the patients who received IVM after doxycycline treatment having no detectable levels of MF, while the remaining patients had very low numbers. This result is in stark contrast to those patients who received IVM alone. These patients had a rise of MF in the skin within 4 months after IVM administration. Importantly, in nodules from patients who had received doxycycline for 6 weeks at 100 mg/day, there was no recrudescence of *Wolbachia* during that time span (although very low levels of DNA were still detectable in a third of the nodules) as determined by immunohistology but also by semiquantitative PCR [88]. This suggests that as long as *Wolbachia* are reduced below a threshold necessary for parasite fertility, the levels will not rebound and make the treatment ineffective in the long run.

Shorter treatments of 4, 3, and 2 weeks administered at 200 mg/day have also been tested. Doxycycline given at 200 mg/day for 4 weeks also blocks embryogenesis and reduces the number of MF in the skin similar to a 6-week regime of 100 mg/day [119]. However, 3 or 2 weeks were insufficient [Hoerauf and Büttner, unpubl. obs.]. Doxycycline treatment for 2 or 3 weeks given 2 months after an initial treatment of 4 or 6 weeks, similar to the regime followed in the *O. ochengi* study [87], failed to kill the adult worms [Hoerauf and Büttner, unpubl. obs.]. Whether this reflects differences between *O. volvulus* and *O. ochengi*, or a difference between the modes of drug administration, i.e. intramuscular injection of a depot formulation of tetracycline for *O. ochengi* versus oral administration for *O. volvulus* needs to be formally tested.

The success of antiwolbachial therapy in *O. volvulus* has also been seen with *W. bancrofti* infections. Again, patients were given 200 mg/day of doxycycline or doxycycline followed by IVM 4 months after commencement of doxycycline administration, IVM or no treatment during the study. All patients received IVM at the conclusion of the study. The therapy was evaluated by measuring MF in the blood and the levels of *Wolbachia* in the MF by quantitative PCR. As seen with onchocerciasis, MF were not detected 12 months after the start of the study in 90% of the patients who received doxycycline and in none of the patients who received IVM after doxycycline. In the IVM alone group, 9% had MF in the blood after 12 months. Copies of the FtsZ gene of the endobacteria were reduced by 96% after 6 weeks of doxycycline therapy [120]. Because the adult worms reside in the lymphatic vessels, it was not possible to examine the embryogenesis in the worms. It is likely that the reduced microfilarial levels are the result of a block in embryogenesis. Thus, antiwolbachial treatment has been shown to be an effective therapy in two of the causative agents of human filariasis.

Probably the most exciting result that has come out of targeting the *Wolbachia* endosymbionts of filarial nematodes with doxycycline is the evidence for adult worm killing [27]. In Tanzania, patients were given 200 mg/day for 8 weeks. As expected, MF levels in the doxycycline-treated patients were reduced to zero after 8 months and remained at this level through the 14-month follow-up in comparison to placebo patients of the same village. Macrofilaricidal activity was measured using a commercially available method that measures antigenemia (worm antigen levels in the blood [121]) and ultrasonography to detect live adult worms at the sites of infection. Ultrasonography is a powerful new tool that is noninvasive and accurate for monitoring macrofilaricidal activity of antifilarial drugs [52, 122, 123]. Antigenemia was significantly reduced 14 months after the start of treatment in patients that received doxycycline. At the 14-month follow-up, 54 adults were examined by ultrasonography for evidence of adult worms. In the doxycycline group, 88% fewer patients were positive for filarial dance sign (adult worm movement) in comparison to the placebo group. Thus, doxycycline depletion of *Wolbachia* from *W. bancrofti* resulted in the death of adult worms. An equivalent macrofilaricidal effect after doxycycline has also been seen in an open study in Ghana where patients received doxycycline for just 6 weeks [118].

Ultrasonography has also been used to document worm killing after DEC treatment in Brazilian and Egyptian patients. In Brazil, worm death was documented in 40% of the patients, much lower than the 88% killing seen after doxycycline treatment [52]. In Egypt, fewer live worms were found following DEC/ALB treatment [124]. The percentage of patients that lost worm nests was similar to that seen with doxycycline. One explanation for the discrepancy between these studies is that the Egyptian study site has had many years of antifilarial chemotherapy and has shown a decline in the number of infections [125]. Because the adult worms cannot be easily taken out of the lymphatic vessels, the condition and age of the worms cannot be determined, making natural attrition the likely reason for the high percentage of adult killing reported in the Egyptian study. The studies with doxycycline [27, 118] and the one in Brazil [52] were both performed in areas where new infections are still occurring, and it can be assumed that younger, fertile worms are present in the treated patients. Despite the occurrence of new infections, patients that received doxycycline had fewer live worms after treatment when compared to the placebo patients and the patients in Brazil that received DEC. Thus, doxycycline has a higher antifilarial efficacy (88% reduction for doxycycline versus 40% for DEC) against adult *W. bancrofti* worms in areas with ongoing transmission.

Recent work has shown that depleting *Wolbachia* from *O. volvulus* also leads to adult worm death (fig. 1a–d) [126]. In this randomized, placebo-controlled study, patients received 200 mg/day of doxycycline or placebo for 6 weeks. All patients received IVM 6 months after the start of the study. Onchocercomas were extirpated at 6, 20, and 27 months after study onset. These were used for quantitative PCR to determine *Wolbachia* depletion, and for immunohistochemistry to determine *Wolbachia* depletion, embryogenesis and fertility, and the proportions of live and dead worms. Doxycycline depleted *Wolbachia* from the nodules by more than 90%, resulting in sterility of adult worms. More significantly, after accounting for acquiring new worms, >70% of the adult worms in the nodules were dead 20 and 27 months after the study onset as shown by staining for aspartic protease, an indicator of worm vitality [127]. This was also reflected in the size of the onchocercomas, which were significantly smaller and fewer in number in patients that received doxycycline. Given the death kinetics of the worms, it could be expected that all worms were killed by 36 months. Worm death was not a result of the development of neoplasms that have been seen after IVM treatment [22, 128] as no worms from doxycycline-treated patients had neoplasms, while 7.6% of the placebo controls did. In contrast to suramin, which is macrofilaricidal, but toxic and requires intravenous administration in a clinical setting [20, 129, 130], antiwolbachial treatment provides the first, low-toxicity method to kill adult *O. volvulus* worms.

Drugs that deplete/kill *Wolbachia* may even help to reverse early stages of LF. An ongoing study of *W. bancrofti*-infected patients in Ghana has shown that MF levels and circulating antigen are significantly reduced in patients that received doxycycline/IVM, but not in placebo/IVM patients. The reduction in antigen is due to adult worm killing as there is a significant decrease in the number of adult worms detected by ultrasonography in doxycycline-treated patients. This study also measured vascular endothelial growth factors (VEGF) and a soluble VEGF receptor (sVEGFR) that are associated with lymphatic vessel dilation. VEGF-C and sVEGFR-3 were significantly reduced in doxycycline/ IVM-treated patients, but not in patients receiving placebo/IVM. A significant reduction was also seen in lymphatic vessel dilation, early stages of lymph edema could be improved in the doxycycline group [131]. Thus, antiwolbachial therapy may also provide the first noninvasive method to treat early stages of LF [58, 131].

Conclusion

Due to its contraindications and the length of time needed to administer, doxycycline is not suitable for MDA. However, doxycycline therapy is a powerful new tool in the battle to eliminate filariasis. Two expert meetings in partnership with the WHO, one in Hamburg [132] and the other in Atlanta (Conference on the Eradicability of Onchocerciasis, Carter Center, January 22–24, 2002) [37] have suggested doxycycline for use in: (a) individuals, e.g. when leaving an endemic area for long periods, and (b) regions which have persistent infections either through remnant foci or resistance. In these cases, a 6-week course of doxycycline to make adult worms sterile would be more cost-effective than continuing annual IVM therapy for another 15 years. Doxycycline is currently the only alternative drug that is approved for human use that can be used against onchocerciasis should the nematodes develop resistance to IVM. Based on the recent data with LF, the indications can be extended for individual therapy, especially when lymphatic pathology such as LE is involved.

An exciting potential of antiwolbachial therapy is the expansion of APOC to selected populations in areas co-endemic for loiasis. The RAPLOA program is a screening method to rapidly identify communities with high *L. loa* MF levels [133, 134]. Currently, such communities are excluded from IVM mass administration. In future, these communities could be treated with doxycycline, or another antiwolbachial drug, to eliminate *Wolbachia* from the *O. volvulus* worms. This would make the *O. volvulus* sterile, halting the production of MF, and lead to adult worm death. As a result, the level of *O. volvulus* MF would slowly decrease through normal attrition of the larvae. Thus, sources of new *O. volvulus* infections could be eliminated despite loiasis co-infection without administering IVM with its associated risk of severe adverse reactions due to the killing of *L. loa* MF. Current studies are examining the feasibility of this use of doxycycline.

New drugs against filariasis are clearly still needed. The results presented here on the efficacy of targeting the *Wolbachia* of filarial nematodes, especially the macrofilaricidal activity seen in human trials with doxycycline, have shown that the endosymbionts are ideal targets for the development of new antifilarial chemotherapies. In addition, there are other benefits to antiwolbachial therapy, i.e. reduction in adverse reactions and even a reduction in early stages of lymphatic pathology, and the possibility to expand filarial control programs into areas co-endemic for *L. loa.* It is hoped, and highly likely, that other drugs already registered for use in humans could be found that have antiwolbachial activity. The search for new drugs which act in a shorter time and that are not contraindicated for a large segment of the community, e.g. pregnant women and children under 8 years of age, is aided by the completion of the sequencing and annotation of the *Wolbachia* genome from *B. malayi* [76, 77].

References

1 WHO: Lymphatic filariasis. WHO Fact Sheet 2000;102.
2 WHO: Onchocerciasis (River Blindness). WHO Fact Sheet 2000;95.
3 WHO: Global programme to eliminate lymphatic filariasis. Wkly Epidemiol Rec 2006;81: 221–232.
4 Basanez MG, Pion SD, Churcher TS, Breitling L, Little MP, Boussinesq M: River blindness: a success story under threat? PLoS Med 2006;3.
5 Gyapong JO, Twum-Danso NA: Editorial: global elimination of lymphatic filariasis: fact or fantasy? Trop Med Int Health 2006;11:125–128.

6 Ottesen EA: Major progress toward eliminating lymphatic filariasis. N Engl J Med 2002;347: 1885–1886.

7 Whitcher JP, Srinivasan M, Upadhyay MP: Corneal blindness: a global perspective. Bull World Health Organ 2001;79:214–221.

8 Kayembe DL, Mwanza JC, Fobi G, et al: Ocular onchocerciasis in seven endemic areas of four Central African countries. Trop Med Int Health 2006, in press.

9 Evans DB, Gelband H, Vlassoff C: Social and economic factors and the control of lymphatic filariasis: a review. Acta Tropica 1993;53:1–26.

10 Vlassoff C, Weiss M, Ovuga EBL, Eneanya C, Newel PT: Gender and the stigma of onchocercal skin disease in Africa. Soc Sci Med 2000;50:1353–1368.

11 Tielsch JM, Beeche A: Impact of ivermectin on illness and disability associated with onchocerciasis. Trop Med Int Health 2004;9:A45–A56.

12 Kirkwood B, Smith P, Marshall T, Prost A: Relationships between mortality, visual acuity and microfilarial load in the area of the Onchocerciasis Control Programme. Trans R Soc Trop Med Hyg 1983;77:862–868.

13 Prost A: The burden of blindness in adult males in the savanna villages of West Africa exposed to onchocerciasis. Trans R Soc Trop Med Hyg 1986;80:525–527.

14 Pion SD, Kamgno J, Demanga N, Boussinesq M: Excess mortality associated with blindness in the onchocerciasis focus of the Mbam Valley, Cameroon. Ann Trop Med Parasitol 2002;96:181–189.

15 Little MP, Breitling LP, Basanez MG, Alley ES, Boatin BA: Association between microfilarial load and excess mortality in onchocerciasis: an epidemiological study. Lancet 2004;363: 1514–1521.

16 WHO: Lymphatic filariasis. Wkly Epidemiol Rec 2001;76:149–154.

17 Omura S, Crump A: The life and times of ivermectin – a success story. Nat Rev Microbiol 2004;2:984–989.

18 Plaisier AP, Alley ES, Boatin BA, Van Oortmarssen GJ, Remme H, De Vlas SJ, Bonneux L, Habbema JD: Irreversible effects of ivermectin on adult parasites in onchocerciasis patients in the Onchocerciasis Control Programme in West Africa. J Infect Dis 1995;172:204–210.

19 Gardon J, Boussinesq M, Kamgno J, Gardon-Wendel N, Demanga N, Duke BO: Effects of standard and high doses of ivermectin on adult worms of *Onchocerca volvulus*: a randomised controlled trial. Lancet 2002;360:203–210.

20 Awadzi K: Clinical picture and outcome of serious adverse events in the treatment of onchocerciasis. Filaria J 2003;2(suppl 1):S6.

21 Awadzi K, Edwards G, Duke BO, Opoku NO, Attah SK, Addy ET, Ardrey AE, Quartey BT: The co-administration of ivermectin and albendazole – safety, pharmacokinetics and efficacy against *Onchocerca volvulus*. Ann Trop Med Parasitol 2003;97:165–178.

22 Duke BO: Evidence for macrofilaricidal activity of ivermectin against female *Onchocerca volvulus*: further analysis of a clinical trial in the Republic of Cameroon indicating two distinct killing mechanisms. Parasitology 2005;130:447–453.

23 Borsboom GJ, Boatin BA, Nagelkerke NJ, Agoua H, Akpoboua KL, Alley EW, Bissan Y, Renz A, Yameogo L, Remme JH, Habbema JD: Impact of ivermectin on onchocerciasis transmission: assessing the empirical evidence that repeated ivermectin mass treatments may lead to elimination/eradication in West-Africa. Filaria J 2003;2:8.

24 Figueredo-Silva J, Jungmann P, Noroes J, Piessens WF, Coutinho A, Brito C, Rocha A, Dreyer G: Histological evidence for adulticidal effect of low doses of diethylcarbamazine in bancroftian filariasis. Trans R Soc Trop Med Hyg 1996;90:192–194.

25 Benton B, Bump J, Seketeli A, Liese B: Partnership and promise: evolution of the African river-blindness campaigns. Ann Trop Med Parasitol 2002;96(suppl 1):S5–S14.

26 Drameh PS, Richards FO, Cross C, Etya'ale DE, Kassalow JS: Ten years of NGDO action against river blindness. Trends Parasitol 2002;18:378–380.

27 Taylor MJ, Makunde WH, McGarry HF, Turner JD, Mand S, Hoerauf A: Macrofilaricidal activity after doxycycline treatment of *Wuchereria bancrofti*: a double-blind, randomised placebo-controlled trial. Lancet 2005;365:2116–2121.

28 Little MP, Basanez MG, Breitling LP, Boatin BA, Alley ES: Incidence of blindness during the Onchocerciasis control programme in western Africa, 1971–2002. J Infect Dis 2004;189:1932–1941.

29 Thylefors B: Eliminating onchocerciasis as a public health problem. Trop Med Int Health 2004;9:A1–A3.
30 Molyneux DH, Bradley M, Hoerauf A, Kyelem D, Taylor MJ: Mass drug treatment for lymphatic filariasis and onchocerciasis. Trends Parasitol 2003;19:516–522.
31 Richards FO Jr, Boatin B, Sauerbrey M, Seketeli A: Control of onchocerciasis today: status and challenges. Trends Parasitol 2001;17:558–563.
32 Molyneux DH, Zagaria N: Lymphatic filariasis elimination: progress in global programme development. Ann Trop Med Parasitol 2002;96(suppl 2):S15–S40.
33 Horton J: Albendazole: a review of anthelmintic efficacy and safety in humans. Parasitology 2000;121(suppl):S113–S132.
34 Addiss D, Critchley J, Ejere H, Garner P, Gelband H, Gamble C: Albendazole for lymphatic filariasis. Cochrane Database Syst Rev 2004:CD003753.
35 Critchley J, Addiss D, Ejere H, Gamble C, Garner P, Gelband H: Albendazole for the control and elimination of lymphatic filariasis: systematic review. Trop Med Int Health 2005;10:818–825.
36 Critchley J, Addiss D, Gamble C, Garner P, Gelband H, Ejere H: Albendazole for lymphatic filariasis. Cochrane Database Syst Rev 2005:CD003753.
37 Dadzie Y, Neira M, Hopkins D: Final report of the Conference on the Eradicability of Onchocerciasis. Filaria J 2003;2:2.
38 Winnen M, Plaisier AP, Alley ES, Nagelkerke NJ, van Oortmarssen G, Boatin BA, Habbema JD: Can ivermectin mass treatments eliminate onchocerciasis in Africa? Bull World Health Organ 2002;80:384–391.
39 Boussinesq M, Gardon J: Prevalences of *Loa loa* microfilaraemia throughout the area endemic for the infection. Ann Trop Med Parasitol 1997;91:573–589.
40 Anonymous: Report of a Scientific Working Group on Serious Adverse Events following Mectizan® treatment of onchocerciasis in *Loa loa* endemic areas. Filaria J 2004;2(suppl 1):S2.
41 Gardon J, Gardon-Wendel N, Demanga N, Kamgno J, Chippaux JP, Boussinesq M: Serious reactions after mass treatment of onchocerciasis with ivermectin in an area endemic for *Loa loa* infection. Lancet 1997;350:18–22.
42 Boussinesq M, Gardon J, Kamgno J, Pion SD, Gardon-Wendel N, Chippaux JP: Relationships between the prevalence and intensity of *Loa loa* infection in the Central province of Cameroon. Ann Trop Med Parasitol 2001;95:495–507.
43 Esterre P, Plichart C, Sechan Y, Nguyen NL: The impact of 34 years of massive DEC chemotherapy on Wuchereria bancrofti infection and transmission: the Maupiti cohort. Trop Med Int Health 2001;6:190–195.
44 Vickery AC, Nayar JK, Tamplin ML: Diethylcarbamazine-mediated clearance of *Brugia pahangi* microfilariae in immunodeficient nude mice. Am J Trop Med Hyg 1985;34:476–483.
45 Maizels RM, Denham DA: Diethylcarbamazine (DEC): immunopharmacological interactions of an anti-filarial drug. Parasitology 1992;105(suppl):S49–S60.
46 McGarry HF, Plant LD, Taylor MJ: Diethylcarbamazine activity against *Brugia malayi* microfilariae is dependent on inducible nitric-oxide synthase and the cyclooxygenase pathway. Filaria J 2005;4:4.
47 Wolstenholme AJ, Rogers AT: Glutamate-gated chloride channels and the mode of action of the avermectin/milbemycin anthelmintics. Parasitology 2005;131(suppl):S85–S95.
48 Vaalburg W, Hendrikse NH, Elsinga PH, Bart J, van Waarde A: P-glycoprotein activity and biological response. Toxicol Appl Pharmacol 2005;207:257–260.
49 Awadzi K, Boakye DA, Edwards G, Opoku NO, Attah SK, Osei-Atweneboana MY, Lazdins-Helds JK, Ardrey AE, Addy ET, Quartey BT, Ahmed K, Boatin BA, Soumbey-Alley EW: An investigation of persistent microfilaridermias despite multiple treatments with ivermectin, in two onchocerciasis-endemic foci in Ghana. Ann Trop Med Parasitol 2004;98:231–249.
50 Awadzi K, Attah SK, Addy ET, Opoku NO, Quartey BT, Lazdins-Helds JK, Ahmed K, Boatin BA, Boakye DA, Edwards G: Thirty-month follow-up of sub-optimal responders to multiple treatments with ivermectin, in two onchocerciasis-endemic foci in Ghana. Ann Trop Med Parasitol 2004;98:359–370.
51 Schwab AE, Boakye DA, Kyelem D, Prichard RK: Detection of benzimidazole resistance-associated mutations in the filarial nematode *Wuchereria bancrofti* and evidence for selection by albendazole and ivermectin combination treatment. Am J Trop Med Hyg 2005;73:234–238.

52 Noroes J, Dreyer G, Santos A, Mendes VG, Medeiros Z, Addiss D: Assessment of the efficacy of diethylcarbamazine on adult *Wuchereria bancrofti* in vivo. Trans R Soc Trop Med Hyg 1997;91:78–81.

53 Eng JK, Prichard RK: A comparison of genetic polymorphism in populations of *Onchocerca volvulus* from untreated- and ivermectin-treated patients. Mol Biochem Parasitol 2005;142:193–202.

54 Ardelli BF, Guerriero SB, Prichard RK: Ivermectin imposes selection pressure on P-glycoprotein from *Onchocerca volvulus*: linkage disequilibrium and genotype diversity. Parasitology 2006;132: 375–386.

55 Hoerauf A, Adjei O, Buttner DW: Antibiotics for the treatment of onchocerciasis and other filarial infections. Curr Opin Investig Drugs 2002;3:533–537.

56 Awadzi K, Edwards G, Opoku NO, Ardrey AE, Favager S, Addy ET, Attah SK, Yamuah LK, Quartey BT: The safety, tolerability and pharmacokinetics of levamisole alone, levamisole plus ivermectin, and levamisole plus albendazole, and their efficacy against *Onchocerca volvulus*. Ann Trop Med Parasitol 2004;98:595–614.

57 Hoerauf A: New strategies to combat filariasis. Expert Rev Anti Infect Ther 2006;4:211–222.

58 Pfarr KM, Hoerauf AM: Antibiotics which target the *Wolbachia* endosymbionts of filarial parasites: a new strategy for control of filariasis and amelioration of pathology. Mini Rev Med Chem 2006;6:203–210.

59 Schares G, Hofmann B, Zahner H: Antifilarial activity of macrocyclic lactones: comparative studies with ivermectin, doramectin, milbemycin A4 oxime, and moxidectin in *Litomosoides carinii, Acanthocheilonema viteae, Brugia malayi,* and *B. pahangi* infection of *Mastomys coucha*. Trop Med Parasitol 1994;45:97–106.

60 Griffin J, Fletcher N, Clemence R, Blanchflower S, Brayden DJ: Selamectin is a potent substrate and inhibitor of human and canine P-glycoprotein. J Vet Pharmacol Ther 2005;28:257–265.

61 Vickers M, Venning M, McKenna PB, Mariadass B: Resistance to macrocyclic lactone anthelmintics by *Haemonchus contortus* and *Ostertagia circumcincta* in sheep in New Zealand. N Z Vet J 2001;49:101–105.

62 Cotreau MM, Warren S, Ryan JL, Fleckenstein L, Vanapalli SR, Brown KR, Rock D, Chen CY, Schwertschlag US: The antiparasitic moxidectin: safety, tolerability, and pharmacokinetics in humans. J Clin Pharmacol 2003;43:1108–1115.

63 Geyer J, Doring B, Godoy JR, Leidolf R, Moritz A, Petzinger E: Frequency of the nt230 (del4) MDR1 mutation in Collies and related dog breeds in Germany. J Vet Pharmacol Ther 2005;28:545–551.

64 Kozek WJ, Figueroa Marroquin H: Intracytoplasmic bacteria in *Onchocerca volvulus*. Am J Trop Med Hyg 1977;26:663–678.

65 McLaren DJ, Worms MJ: Micro-organisms in filarial larvae (Nematoda). Trans R Soc Trop Med Hyg 1975;69:509–514.

66 Kozek WJ: Transovarially-transmitted intracellular microorganisms in adult and larval stages of *Brugia malayi*. J Parasitol 1977;63:992–1000.

67 Kozek WJ, Orihel TC: Ultrastructure of *Loa loa* microfilaria. Int J Parasitol 1983;13:19–43.

68 Taylor MJ, Hoerauf A: *Wolbachia* bacteria of filarial nematodes. Parasitol Today 1999;15: 437–442.

69 Hoerauf A, Büttner DW, Adjei O, Pearlman E: Onchocerciasis. BMJ 2003;326:207–210.

70 Sironi M, Bandi C, Sacchi L, Di Sacco B, Damiani G, Genchi C: Molecular evidence for a close relative of the arthropod endosymbiont *Wolbachia* in a filarial worm. Mol Biochem Parasitol 1995;74:223–227.

71 Taylor MJ, Bandi C, Hoerauf A: *Wolbachia* bacterial endosymbionts of filarial nematodes. Adv Parasitol 2005;60:245–284.

72 Bandi C, Anderson TJC, Genchi C, Blaxter ML: Phylogeny of *Wolbachia* in filarial nematodes. Proc R Soc Lond 1998;265:2407–2413.

73 Lo N, Casiraghi M, Salati E, Bazzocchi C, Bandi C: How many *Wolbachia* supergroups exist? Mol Biol Evol 2002;19:341–346.

74 Jiggins FM: The rate of recombination in *Wolbachia* bacteria. Mol Biol Evol 2002;19:1640–1643.

75 Jiggins FM, Hurst GDD, Yang Z: Host-symbiont conflicts: positive selection on an outer membrane protein of parasitic but not mutualistic Rickettsiaceae. Mol Biol Evol 2002;19:1341–1349.

76 Foster J, Ganatra M, Kamal I, Ware J, Makarova K, Ivanova N, Bhattacharyya A, Kapatral V, Kumar S, Posfai J, Vincze T, Ingram J, Moran L, Lapidus A, Omelchenko M, Kyripide N, Ghedin E, Wang S, Goltsman E, Joukov V, Ostravskaya O, Tsukerman K, Mazur M, Comb D, Koonin E,

Slatko B: The *Wolbachia* genome of *Brugia malayi*: endosymbiont evolution within a human pathogenic nematode. PLoS Biol 2005;3:e121.

77 Pfarr K, Hoerauf A: The annotated genome of *Wolbachia* from the filarial nematode *Brugia malayi*: what it means for progress in antifilarial medicine. PLoS Med 2005;2:e110.

78 Büttner DW, Wanji S, Bazzocchi C, Bain O, Fischer P: Obligatory symbiotic *Wolbachia* endobacteria are absent from *Loa loa*. Filaria J 2003;2:10.

79 Grobusch MP, Kombila M, Autenrieth I, Mehlhorn H, Kremsner PG: No evidence of *Wolbachia* endosymbiosis with *Loa loa* and *Mansonella perstans*. Parasitol Res 2003;90:405–408.

80 McGarry HF, Pfarr K, Egerton G, Hoerauf A, Akue JP, Enyong P, Wanji S, Klager SL, Bianco AE, Beeching NJ, Taylor MJ: Evidence against *Wolbachia* symbiosis in *Loa loa*. Filaria J 2003;2:9.

81 Casiraghi M, Favia G, Cancrini G, Bartoloni A, Bandi C: Molecular identification of *Wolbachia* from the filarial nematode *Mansonella ozzardi*. Parasitol Res 2001;87:417–420.

82 Taylor MJ, Hoerauf A: A new approach to the treatment of filariasis. Curr Opin Infect Dis 2001;14:727–737.

83 Genchi C, Sacchi L, Bandi C, Venco L: Preliminary results on the effect of tetracycline on the embryogenesis and symbiotic bacteria (*Wolbachia*) of *Dirofilaria immitis*. An update and discussion. Parassitologia 1998;40:247–249.

84 Bandi C, McCall JW, Genchi C, Corona S, Venco L, Sacchi L: Effects of tetracycline on the filarial worms *Brugia pahangi* and *Dirofilaria immitis* and their bacterial endosymbionts *Wolbachia*. Inter J Parasitol 1999;29:357–364.

85 Hoerauf A, Nissen-Pähle K, Schmetz C, Henkle-Dührsen K, Blaxter ML, Büttner DW, Gallin MY, Al-Qaoud KM, Lucius R, Fleischer B: Tetracycline therapy targets intracellular bacteria in the filarial nematode *Litomosoides sigmodontis* and results in filarial infertility. J Clin Invest 1999;103:11–18.

86 Hoerauf A, Volkmann L, Nissen-Pähle K, Schmetz C, Autenrieth I, Büttner DW, Fleischer B: Targeting of *Wolbachia* endobacteria in *Litomosoides sigmodontis*: comparison of tetracyclines with chloramphenicol, macrolides and ciproflaxcin. Trop Med Inter Health 2000;5:275–279.

87 Langworthy NG, Renz A, Mackenstedt U, Henkle-Dührsen K, de C Bronsvoort MB, Tanya VN, Donnelly MJ, Trees AJ: Macrofilaricidal activity of tetracycline against the filarial nematode *Onchocerca ochengi*: elimination of *Wolbachia* precedes worm death and suggests a dependent relationship. Proc Royal Soc Lond B 2000;267:1063–1069.

88 Hoerauf A, Mand S, Volkmann L, Buttner M, Marfo-Debrekyei Y, Taylor M, Adjei O, Buttner DW: Doxycycline in the treatment of human onchocerciasis: kinetics of *Wolbachia* endobacteria reduction and of inhibition of embryogenesis in female *Onchocerca* worms. Microbes Infect 2003;5:261–273.

89 Volkmann L, Fischer K, Taylor M, Hoerauf A: Antibiotic therapy in murine filariasis (*Litomosoides sigmodontis*): comparative effects of doxycycline and rifampicin on *Wolbachia* and filarial viability. Trop Med Int Health 2003;8:392–401.

90 Gilbert J, Nfon CK, Makepeace BL, Njongmeta LM, Hastings IM, Pfarr KM, Renz A, Tanya VN, Trees AJ: Antibiotic chemotherapy of onchocerciasis: in a bovine model, killing of adult parasites requires a sustained depletion of endosymbiotic bacteria (*Wolbachia* species). J Infect Dis 2005;192:1483–1493.

91 Rao R, Moussa H, Vanderwaal RP, Sampson E, Atkinson LJ, Weil GJ: Effects of gamma radiation on Brugia malayi infective larvae and their intracellular *Wolbachia* bacteria. Parasitol Res 2005;97:219–227.

92 Smith HL, Rajan TV: Tetracycline inhibits development of the infective-stage larvae of filarial nematodes in vitro. Exp Parasitol 2000;95:265–270.

93 Bosshardt SC, McCall JW, Coleman SU, Jones KL, Petit TA, Klei TR: Prophylactic activity of tetracycline against *Brugia pahangi* infection in jirds (*Meriones unquiculatus*). J Parasitol 1993;79: 775–777.

94 Rajan TV: Relationship of anti-microbial activity of tetracyclines to their ability to block the L3 to L4 molt of the human filarial parasite *Brugia malayi*. Am J Trop Med Hyg 2004;71:24–28.

95 Bazzocchi C, Ceciliani F, McCall JW, Ricci I, Genchi C, Bandi C: Antigenic role of the endosymbionts of filarial nematodes: IgG response against the *Wolbachia* surface protein in cats infected with *Dirofilaria immitis*. Proc R Soc Lond 2000;267:2511–2516.

96 Punkosdy GA, Dennis VA, Lasater BL, Tzertzinis G, Foster JM, Lammie PJ: Detection of serum IgG antibodies specific for *Wolbachia* surface protein in Rhesus monkeys infected with *Brugia malayi*. J Infect Dis 2001;184:385–389.

97 Saint André A, Blackwell NM, Hall LR, Hoerauf A, Brattig NW, Volkmann L, Taylor MJ, Ford L, Hise AG, Lass JH, Diaconu E, Pearlman E: The role of endosymbiotic *Wolbachia* bacteria in the pathogenesis of river blindness. Science 2002;295:1892–1895.

98 Hise AG, Gillette-Ferguson I, Pearlman E: Immunopathogenesis of *Onchocerca volvulus* keratitis (river blindness): a novel role for TLR4 and endosymbiotic *Wolbachia* bacteria. J Endotoxin Res 2003;9:390–394.

99 Akira S: Toll-like receptor signaling. J Biol Chem 2003;278:38105–38108.

100 Gillette-Ferguson I, Hise AG, Sun Y, Diaconu E, McGarry HF, Taylor MJ, Pearlman E: *Wolbachia*- and *Onchocerca volvulus*-induced keratitis (river blindness) is dependent on myeloid differentiation factor 88. Infect Immun 2006;74:2442–2445.

101 Brattig NW, Rathjens U, Ernst M, Geisinger F, Renz A, Tischendorf FW: Lipopolysaccharide-like molecules derived from *Wolbachia* endobacteria of the filaria *Onchocerca volvulus* are candidate mediators in the sequence of inflammatory and antiinflammatory responses of human monocytes. Microbes Infect 2000;2:1147–1157.

102 Taylor MJ, Cross HF, Bilo K: Inflammatory responses induced by the filarial nematode *Brugia malayi* are mediated by lipopolysaccharide-like activity from endosymbiotic *Wolbachia* bacteria. J Exp Med 2000;191:1429–1436.

103 O'Neill SL, Pettigrew MM, Sinkins SP, Braig HR, Andreadis TG, Tesh RB: In vitro cultivation of *Wolbachia pipientis* in an *Aedes albopictus* cell line. Insect Mol Biol 1997;6:33–39.

104 Brattig NW, Büttner DW, Hoerauf A: Neutrophil accumulation around *Onchocerca* worms and chemotaxis of neutrophils are dependent on *Wolbachia* endobacteria. Microbes Infect 2001;3:439–446.

105 Brattig NW, Bazzocchi C, Kirschning CJ, Reiling N, Büttner DW, Ceciliani F, Geisinger F, Hochrein H, Ernst M, Wagner H, Bandi C, Hoerauf A: The major surface protein of *Wolbachia* endosymbionts in filarial nematodes elicits responses through TLR2 and TLR4. J Immunol 2004;173:437–445.

106 Boreham PFL, Atwell RB: Adverse drug reactions in the treatment of filarial parastites: hematological, biochemical, immunological and pharmacological changes in *Dirofilaria immitis* infected dogs treated with diethylcarbamazine. Journal Long Form workform 1983;13:547–556.

107 Francis H, Awadzi K, Ottesen EA: The Mazzotti reaction following treatment of *Onchocerciasis* with diethylcarbamazine: clinical severity as a function of infection intensity. Am J Trop Med Hyg 1985;34:529–536.

108 Haarbrink M, Abadi GK, Buurman WA, Dentener MA, Terhell AJ, Yazdanbakhsh M: Strong association of interleukin-6 and lipopolysaccharide-binding protein with severity of adverse reactions after diethylcarbamazine treatment of microfilaremic patients. J Infect Dis 2000;182:564–569.

109 Babu BV, Rath K, Kerketta AS, Swain BK, Mishra S, Kar SK: Adverse reactions following mass drug administration during the Programme to Eliminate Lymphatic Filariasis in Orissa State, India. Trans R Soc Trop Med Hyg 2006;100:464–469.

110 Cross HF, Haarbrink M, Egerton G, Yazdanbakhsh M, Taylor MJ: Severe reactions to filarial chemotherapy and release of *Wolbachia* endosymbionts into blood. Lancet 2001;358:1873–1875.

111 Turner JD, Mand S, Debrah AY, Muehlfeld J, Pfarr K, McGarry HF, Adjei O, Taylor MJ, Hoerauf A: A randomized, double-blind clinical trial of a 3-week course of doxycycline plus albendazole and ivermectin for the treatment of *Wuchereria bancrofti* infection. Clin Infect Dis 2006;42: 1081–1089.

112 Keiser PB, Reynolds SM, Awadzi K, Ottesen EA, Taylor MJ, Nutman TB: Bacterial endosymbionts of *Onchocerca volvulus* in the pathogenesis of posttreatment reactions. J Infect Dis 2002;185:805–811.

113 Stanley SL Jr, Kell O: Ascending paralysis associated with diethylcarbamazine treatment of M. *loa loa* infection. Trop Doct 1982;12:16–19.

114 Carme B, Boulesteix J, Boutes H, Puruehnce MF: Five cases of encephalitis during treatment of loiasis with diethylcarbamazine. Am J Trop Med Hyg 1991;44:684–690.

115 Godoy GA: Circulating immune complexes in *Mansonella ozzardi* infection. Ann Trop Med Parasitol 1998;92:895–896.

116 Bartoloni A, Cancrini G, Bartalesi F, Marcolin D, Roselli M, Arce CC, Hall AJ: *Mansonella ozzardi* infection in Bolivia: prevalence and clinical associations in the Chaco region. Am J Trop Med Hyg 1999;61:830–833.

117 Taylor MJ, Makunde WH, McGarry HF, Mand S, Hoerauf A: Doxycycline treatment of *Wuchereria bancrofti*: a double-blind placebo-controlled trial. Am J Trop Med Hyg 2003;69:S250.
118 Debrah A, Mand S, Specht S, Batsa L, Marfo Y, Pfarr K, Larbi J, Adjei O, Hoerauf A: Effect of targeting endosymbiotic *Wolbchia* in *Wuchereria bancrofti* on macrofilaricidal effects, overexpression of VEGF-C/VEGFR-3 and lymphatic dilation in lymphatic filariasis-results from a 6 weeks trial with doxycycline. Am J Trop Med Hyg 2005;73(suppl):abstract 2336.
119 Debrah AY, Mand S, Marfo-Debrekyei Y, Larbi J, Adjei O, Hoerauf A: Assessment of microfilarial loads in the skin of onchocerciasis patients after treatment with different regimens of doxycycline plus ivermectin. Filaria J 2006;5:1.
120 Hoerauf A, Mand S, Fischer K, Kruppa T, Marfo-Debrekyei Y, Debrah Alexander Y, Pfarr KM, Adjei O, Büttner DW: Doxycycline as a novel strategy against bancroftian filariasis-depletion of *Wolbachia* endosymbionts from *Wuchereria bancrofti* and stop of microfilariae production. Med Microbiol Immunol (Berl) 2003;5:261–273.
121 Simonsen PE, Dunyo SK: Comparative evaluation of three new tools for diagnosis of bancroftian filariasis based on detection of specific circulating antigens. Trans R Soc Trop Med Hyg 1999;93:278–282.
122 Dreyer G, Amaral F, Noroes J, Medeiros Z: Ultrasonographic evidence for stability of adult worm location in bancroftian filariasis. Trans R Soc Trop Med Hyg 1994;88:558.
123 Mand S, Marfo-Debrekyei Y, Dittrich M, Fischer K, Adjei O, Hoerauf A: Animated documentation of the filaria dance sign (FDS) in bancroftian filariasis. Filaria J 2003;2:3.
124 Hussein O, Setouhy ME, Ahmed ES, Kandil AM, Ramzy RM, Helmy H, Weil GJ: Duplex Doppler sonographic assessment of the effects of diethylcarbamazine and albendazole therapy on adult filarial worms and adjacent host tissues in Bancroftian filariasis. Am J Trop Med Hyg 2004;71: 471–477.
125 Ramzy RM, El Setouhy M, Helmy H, Ahmed ES, Abd Elaziz KM, Farid HA, Shannon WD, Weil GJ: Effect of yearly mass drug administration with diethylcarbamazine and albendazole on bancroftian filariasis in Egypt: a comprehensive assessment. Lancet 2006;367:992–999.
126 Hoerauf A, Specht S, Büttner M, Pfarr KM, Mand S, Debrekyei YM, Konadu P, Debrah AY, Bandi C, Brattig N, Albers A, Larbi J, Basta L, Adjei O, Büttner DW: Macrofilaricidal activity and *Wolbachia* depletion by doxycycline in onchocerciasis. PLoS Med, submitted.
127 Jolodar A, Fischer P, Buttner DW, Miller DJ, Schmetz C, Brattig NW: *Onchocerca volvulus*: expression and immunolocalization of a nematode cathepsin D-like lysosomal aspartic protease. Exp Parasitol 2004;107:145–156.
128 Duke BO, Marty AM, Peett DL, Gardo J, Pion SD, Kamgno J, Boussinesq M: Neoplastic change in *Onchocerca volvulus* and its relation to ivermectin treatment. Parasitology 2002;125:431–444.
129 Duke BO: The effects of drugs on *Onchocerca volvulus*. 3. Trials of suramin at different dosages and a comparison of the brands Antrypol, Moranyl and Naganol. Bull World Health Organ 1968;39:157–167.
130 Awadzi K, Hero M, Opoku NO, Addy ET, Buttner DW, Ginger CD: The chemotherapy of onchocerciasis XVIII. Aspects of treatment with suramin. Trop Med Parasitol 1995;46:19–26.
131 Debrah AY, Mand S, Specht S, Marfo-Debrekyei Y, Batsa L, Pfarr K, Larbi J, Lawson B, Taylor M, Adjei O, Hoerauf A: Doxycycline reduces plasma VEGF-C/sVEGFR-3 and improves pathology in lymphatic filariasis. PLoS Pathog 2006;2:e92.
132 Hoerauf A, Fleischer B, Walter RD: Of filariasis, mice and men. Trends Parasitol 2001;17:4–5.
133 Addiss DG, Rheingans R, Twum-Danso NA, Richards FO: A framework for decision-making for mass distribution of Mectizan® in areas endemic for *Loa loa*. Filaria J 2003;2(suppl 1):S9.
134 Wanji S, Tendongfor N, Esum M, Yundze SS, Taylor MJ, Enyong P: Combined utilisation of rapid assessment procedures for loiasis (RAPLOA) and onchocerciasis (REA) in rain forest villages of Cameroon. Filaria J 2005;4:2.

Kenneth Pfarr
Institute for Medical Microbiology, Immunology and Parasitology
University Clinic Bonn, Sigmund-Freud-Strasse 25
DE–53105 Bonn (Germany)
Tel. +49 228 287 11510, Fax +49 228 287 14330, E-Mail pfarr@parasit.meb.uni-bonn.de

Hoerauf A, Rao RU (eds): Wolbachia.
Issues Infect Dis. Basel, Karger, 2007, vol 5, pp 52–65

It Takes Two: Lessons From the First Nematode *Wolbachia* Genome Sequence

Kenneth Pfarr[a], *Jeremy Foster*[b], *Barton Slatko*[b]

[a]University Clinic Bonn, Institute for Medical Microbiology, Immunology and Parasitology, Bonn, Germany; [b]Molecular Parasitology Division, New England Biolabs, Ipswich, Mass., USA

Abstract

Lymphatic filariasis and onchocerciasis are major agents of morbidity in developing countries of the tropics. While current control methods have proven successful in different countries, the strategies require treatment times spanning decades. Of greater concern is the dependence on a single drug therapy for onchocerciasis for which there is evidence of resistance. *Wolbachia* are attractive targets for control of filariasis as the endosymbionts appear to be obligate for human filarial nematode vitality. Antibiotic studies in both animal models and in human trials indicate that disruption of *Wolbachia* leads to serious negative consequences for nematode development and reproduction. Annotation of the genome sequence of the *Wolbachia* endosymbiont from *Brugia malayi* suggests biochemical pathways utilized in the host-symbiont interaction that are potential targets for inhibiting the nematode life cycle. Possibilities include provision of nucleotides and heme by *Wolbachia* to the nematode host and conversely, provision of amino acids by the host nematode to the *Wolbachia*. Current genome initiatives are overcoming difficulties in purification of *Wolbachia* DNAs from host genomic DNAs that have inhibited rapid sequencing of other *Wolbachia*. Comparative analysis from these genomes will be useful in determining the underlying biology of the host-symbiont relationship and help further elucidate pathways for antiwolbachial targeting.

*'Great fleas have little fleas upon their backs, to bite 'em.
And little fleas have lesser fleas and so on, ad infinitum.'*

Augustus DeMorgan, 1806–1871
(based upon a poem of Jonathan Swift, 1667–1745)

Over the last several years, it has become clear that *Wolbachia* endosymbionts may provide biochemical targets for control of human filarial diseases. These maladies affect 150 million people worldwide, with over 1 billion people in more than 90 countries at risk from the insect-borne parasitic nematodes. The nematodes are responsible for lymphatic or cutaneous filariasis, leading to medical conditions including elephantiasis or onchocerciasis (African river blindness). Lymphatic filariasis is caused predominantly by *Wuchereria bancrofti* and *Brugia malayi* and affects 120 million individuals, a third of whom show disfigurement. Onchocerciasis, caused by *Onchocerca volvulus*, affects 18 million people, of whom 500,000 have visual impairment and 270,000 are blind [1, 2].

Almost 30 years ago, *Wolbachia* intracellular bacteria were observed within particular tissues of these filarial parasites by electron microscopy, although they were not identified at that time [3–6]. During the Filarial Genome Project directed towards *B. malayi*, the presence of rare α-proteobacterial cDNA sequences among those expected from *B. malayi* suggested the occurrence of endobacterial DNA [7]. While the cDNA library construction utilized poly-T primers to initiate first strand cDNA synthesis from polyA+-containing transcripts, priming occasionally occurred off of the A+U rich *Wolbachia* RNA, producing rare cDNA clones of endosymbiont origin.

The endobacterial sequences were subsequently identified as *Wolbachia* by phylogenetic analysis [8]. *Wolbachia* endosymbionts can be separated into six supergroups based upon 16S rRNA, *Wolbachia* surface protein (WSP), *gro*EL, *glt*A and *fts*Z phylogenetics [8–17]. Four supergroups contain *Wolbachia* from arthropods, while supergroup C contains *Wolbachia* from the nematodes *O. volvulus* and *Dirofilaria immitis* (canine and feline hosts), and supergroup D contains *Wolbachia* from *B. malayi*, *W. bancrofti*, and *Litomosoides sigmodontis* (cotton rat host) [9, 16]. In nematodes, the evolution of *Wolbachia* parallels the phylogenetics of their hosts, while in the other supergroups, horizontal transmission appears to have occurred [9, 10, 13, 16, 18].

Wolbachia endosymbionts have now been found in the vast majority of filarial nematode species [3, 9, 10, 18–26]. Among the subfamilies Onchocercinae and Dirofilariinae, *Wolbachia* occur in the main agents of human and animal filariasis. Exceptions among the filariae of humans are *Loa loa* and *Mansonella perstan* [25–28].

In nematodes which contain *Wolbachia* and which have been well examined, the bacteria are present in all life cycle stages of the nematode hosts and are located in both sexes in the hypodermal cells within the lateral chords (invaginations of the body wall hypodermis that project into the body cavity). They are also localized in ovaries, oocytes and developing embryonic stages within female uteri but not in the male reproductive system, suggesting that the bacterium is vertically transmitted through the cytoplasm of the egg and not through sperm [5, 29].

Within species of nematodes containing *Wolbachia*, all nematodes contain *Wolbachia*, suggesting that they are required for worm fertility and survival [20, 30, 31]. The *Wolbachia* vary in their proportions within life cycle stages [5, 32, 33], suggesting differential endobacterial growth rates among the filarial worm life cycle stages. *Wolbachia* numbers remain constant in microfilariae and the mosquito-borne larval stages (L2 and L3), but, in contrast, bacterial numbers increase dramatically during the first week of host infection. The increase continues throughout L4 larval development. In females, bacteria numbers continue to increase as the ovary and embryonic larval stages become infected [32].

As discussed in other chapters of this book, *Wolbachia* also play a significant role in the host immunological response to filarial parasite invasion, including induction of antibodies directed toward *Wolbachia*-specific antigens, such as WSP, heat shock protein, aspartate aminotransferase and Htr serine protease [9, 34–39]. Release of filarial worm-associated molecules, especially after drug treatments that cause worm death in the host, leads to pathogenesis ('Mazzotti reaction') [40–44], and *Wolbachia* has been associated with chronic and acute infection states of filariasis [reviewed in 45].

Wolbachia products induce potent inflammatory activity [46, 47]. The inflammatory activity of lymphatic filarial worms is induced in part by repeated exposure to *Wolbachia*-mediated inflammation following death of filariae [38, 48–50]. A recent finding showed that *Wolbachia* numbers are more abundant in *O. volvulus* sampled from infections where severe ocular disease is common, compared to samples from a forested area where blindness is rare [51]. One implication may be that *Wolbachia* boosts immune responsiveness toward filarial antigens, facilitating the clearance of microfilariae and the development of immunopathogenesis.

The ubiquious presence of *Wolbachia* in filarial nematodes harboring them suggested a critical role in nematode development and reproduction. Evidence for an obligatory dependence of *Wolbachia* in the nematode host has been demonstrated using antibiotics such as doxycycline, tetracycline, rifampicin and azithromycin, which have detrimental effects on filarial nematodes containing *Wolbachia* (nematode female sterility, worm lethality), whereas there were no observed effects on *Wolbachia*-free filariae [22, 52–58].

Recent human drug trials with doxycycline resulted in essential depletion of *Wolbachia*, correlating with a phenotypic block in embryogenesis, a reduced microfilarial output and long-term sterilization of *O. volvulus* and macrofilaricidal effects in *W. bancrofti* [56, 59–61] using 6- to 8-week courses of doxycycline. The importance of *Wolbachia* for adult survival was demonstrated by the reduction in circulating filarial antigen, the current gold standard for monitoring the loss of *W. bancrofti* worms, and the reduction in the number of worm nests detected by ultrasonography, a noninvasive method for determining worm loss

[56], in the groin area of infected patients. Antibiotic treatment for relatively long periods of time largely eliminates *Wolbachia*, reduces transmission of filarial nematodes and eliminates the pathogenic life stages of onchocerciasis and lymphatic filariasis. However, the length of treatment, contraindications of doxycycline (not recommended for children under 9 or for pregnant or breast-feeding women), and the potential for drug resistance, argue for a need to identify additional antiwolbachial agents that can be given to everyone in endemic areas as part of the current mass drug administration programs outlined in the chapter by Hoerauf and Pfarr [pp 31 –51].

As detailed in the chapter on the efficacy of antiwolbachial therapy in the battle against filarial infections, previous strategies for elimination of filariasis have included vector control in the presence or absence of antiparasitic drugs [62–66]. Diethylcarbamazine, albendazole, and ivermectin have been the most recent drugs of choice for prevention of filarial infections, but since they have little effect on adult worms, repeated doses in endemic areas are required to eliminate infections that can arise again within months of treatment [67–69]. In addition, the possibility of drug resistance, as observed with intestinal helminths in animals is a concern [70, 71]. Other than doxycycline, no new therapeutics have been developed in over 20 years. For the reasons given above, there is a need for better drugs that permanently sterilize or kill adult worms. Targeting *Wolbachia* may fulfill that need. One tool that could help in identifying new therapeutics is the genome sequence of the *Wolbachia* from filarial nematodes.

Genomic sequencing and annotation of the *Wolbachia* endosymbiont from *B. malayi* (*w*Bm) was undertaken to better understand the biology of *Wolbachia* and its interaction with the nematode host [72, 73]. The *w*Bm genome is 1.1 Mb in length and is 66% A+T in composition, similar to the A+T content determined for the DNA of the nematode host. Annotation pinpoints 806 predicted protein coding genes, and 696 *w*Bm proteins have an ortholog in the only other completed *Wolbachia* genome, that of *Drosophila melanogaster* (*w*Mel) [74].

Comparative analysis of other *Wolbachia* genomes will help to further pinpoint common biochemical pathways for intervention, but technical challenges of purifying DNA from obligate endosymbionts of the genus *Wolbachia* have hindered studies aimed at characterizing and sequencing their genomes. In the case of the *Wolbachia* present in filarial nematodes, these problems are confounded by the limited availability of biological tissue from which to attempt purification of the bacterial DNA. Many filarial nematodes of medical importance, such as *W. bancrofti* and *O. volvulus*, are restricted to the tropics and lack a laboratory host system for maintenance of the life cycle. Only the human parasite *B. malayi* and *L. sigmodontis* are readily maintained in the laboratory. While *D. immitis* is found throughout the world, it is usually only recovered from dogs postmortem or by a complicated surgical procedure.

Despite these limitations, considerable advances have been made in purifying and chararacterizing *Wolbachia* genomes as a first step towards unraveling the biological complexities of these fascinating endosymbionts. The complete *Wolbachia* genomes from one strain of *D. melanogaster* (*w*Mel) and from the nematode *B. malayi* (*w*Bm) have now been reported [72, 74] and many more are in the pipeline [75].

Pulsed field gel electrophoresis (PFGE) has greatly facilitated studies of *Wolbachia* genomes. Initially, the sizes of *Wolbachia* genomes from different strains of *D. melanogaster* and two filarial nematodes were determined by PFGE of the intact genome and of DNA fragments generated using rare-cutting restriction endonucleases [76]. It was noted that the *Wolbachia* genomes recovered from nematode hosts were around 1 Mb while those from the insect hosts were larger, ranging from about 1.35–1.65 Mb. These studies paved a way for purification of *Wolbachia* DNA for whole genome sequencing. Unfortunately, the large circular bacterial chromosomes were observed to enter the agarose pulse field gels very poorly with most of the DNA remaining in the wells along with the high molecular weight host organism DNA. It was presumed that only nicked copies of the *Wolbachia* chromosome entered the agarose matrix [76]. However, in the case of the *Drosophila* endosymbiont, *Wolbachia* DNA was recovered from pulsed field gels in amounts sufficient for full genomic sequencing following restriction digestion with AscI to cleave the circular genome into two fragments [74]. Since the available *Wolbachia* genome sequences contain only one copy of the 23S rRNA gene that typically contains a restriction site for the extremely rare-cutting I-Ceu I [77], this enzyme could find future application in linearizing *Wolbachia* genomes prior to PFG purification. Similarly, the use of a nicking enzyme to change the conformation of the intact circular chromosome to aid its entrance into the agarose matrix remains unexplored. For sequencing *w*Bm, too little nematode material was available to make such a PFG purification strategy feasible, but again PFGE proved invaluable. A bacterial artificial chromosome library with average insert sizes of 50 kb was prepared from *B. malayi* genomic DNA which naturally contains a low level of *Wolbachia* DNA. Clones containing endosymbiont DNA were identified by hybridization until a contig spanning the entire genome was assembled [73] and the clones were subsequently sequenced to determine the complete genome [72]. Although time consuming and labor intensive, this strategy has several advantages over the whole genome shotgun sequencing strategy used for the *Drosophila* endosymbiont. Firstly, there is no requirement for large amounts of *Wolbachia* DNA. Second, despite PFG purification, almost 40% of the sequence contigs initially assembled for *w*Mel derived from *Drosophila* DNA not *Wolbachia* [74], decreasing the cost efficiency of sequencing. Third, the highly repetitive nature of *Wolbachia* DNA complicated sequence assembly in the whole genome shotgun approach [74]. In a BAC-based sequencing strategy, no host organism DNA is present and by sequencing the genome in distinct pieces (BAC clone inserts) assembly is greatly simplified.

A clone-based sequencing strategy has also been initiated for the *Wolbachia* endosymbiont of *O. volvulus* [78] as direct DNA sequencing of that genome has proven to be difficult due to problems of obtaining *Wolbachia* DNA [Fenn and Whitton, pers. commun.]. However, similar to the presence of *w*Bm in genomic BAC libraries from *B. malayi*, several *w*Ovo clones were identified from an *O. volvulus* large insert lambda Fix DNA library (9- to 23-kb inserts) by using previously known *w*Ovo EST sequences and other related sequences (*wsp*, 16S, 23S ribosomal RNA genes, *GroEL*, etc.) as probes. About 71 kb was sequenced, which provides about 6.5% nonredundant *w*Ovo genome sequence. Comparison of the genome organization of the *w*Ovo fragments with *w*Mel and *w*Bm shows large genome rearrangements. In fact, the genome organization of *w*Mel and *w*Bm is much more similar to each other than either are to *w*Ovo in four out of the five compared *w*Ovo fragments [Fenn et al., pers. commun.]. The lack of synteny observed in *Wolbachia* genomes, even when the *Wolbachia* originate from host organisms of the same genus, for example *Drosophila* [79–81], effectively eliminates 'walking', 'skim sequencing' or long-range PCR strategies for completing the genome sequence based upon syntenic genome structure.

Certain other *Wolbachia* genomes have been sequenced inadvertently by virtue of their DNA being present in the libraries made for whole genome sequencing of their host organism. This was true for *w*Bm [72] where the genome was independently sequenced as part of the sequencing project for *B. malayi* [82] and also is the case for various *Wolbachia* genomes from different species of *Drosophila* [79–81]. While some other *Wolbachia* genomes may be sequenced from purified genomic DNA (for example the *Culex Wolbachia* genome [83], this approach has generally proven to be technically difficult. For example, the *Wolbachia* genome from *D. immitis* is similarly being sequenced from purified DNA but, once again, while DNA purification has been difficult, only limited sequence has been currently obtained [Bandi, pers. commun.].

An exciting new method for amplifying the *Wolbachia* genome from small amounts of pulsed field gel-purified DNA has been described [84]. Up to 10 μg *Wolbachia* DNA was produced by multiple-displacement amplification of as little as 1 ng purified template DNA. The *w*Ri strain of *Wolbachia* from *Drosophila simulans* was amplified and all loci that were subsequently targeted by PCR were represented, suggesting that all or most of the genome had been amplified.

As mentioned previously, the *w*Bm genome was completed in a clone-based approach due to difficulty of purification of large enough amounts of

DNA for shotgun sequencing approaches. The annotation analysis [72] reveals the basic biochemistry of *w*Bm and provides information on potential targets for biological intervention. The annotation suggests that pyruvate and Krebs cycle intermediates derived from amino acids are utilized in gluconeogenesis, rather than glycolysis. A pyruvate dehydrogenase complex, a complete Krebs cycle and respiratory chain elements typical of α-proteobacteria are present, as are numerous proteases and peptidases that likely degrade host proteins in the extracellular environment.

wBm may contribute riboflavin, flavin adenine dinucleotide, heme and nucleotides to the symbiotic relationship with *B. malayi*. *w*Bm contains all enzymes for riboflavin and flavin adenine dinucleotide biosynthesis and has complete pathways for de novo synthesis of purines and pyrimidines. *w*Bm may be an essential source of nucleotides for the host, especially during embryogenesis where the nucleotide requirement may be high. In the host *B. malayi* genome sequence, the purine metabolism genetic pathway appears to be absent [Ghedin, pers. commun.]. Rajan [85] has shown that the in vitro L3–L4 molt requires exogenous nucleosides.

Conversely, the nematode host in this symbiotic relationship is likely providing amino acids required for the endobacterial growth, since *w*Bm can only synthesize one amino acid, meso-diaminopimelate, a major component of peptidoglycan. The cell wall biosynthesis pathways are devoid of genes required for the biosynthesis of lipopolysaccharide, similar to *w*Mel [74]. *w*Bm likely makes unmodified peptidoglycan, while *w*Mel has retained genes that can modify peptidoglycan with oligosaccharide. Differences in peptidoglycan structure between *w*Bm and *w*Mel suggest adaptations to their respective mutualistic or parasitic lifestyles and might be interesting drug targets. *w*Bm and *w*Mel lack many genes for synthesis of lipid A, the usual component of proteobacterial membranes [72, 74].

Ankyrin domain-containing proteins are of interest because of their roles involving protein-protein interactions in a variety of cellular processes. Of the 12 ankyrin genes in *w*Bm, 7 are pseudogenes and of the remaining 5, at least 4 are expressed as evidenced by RT-PCR and microarray experiments [Ware et al., unpubl.; Scott and Slatko, unpubl.]. Ankyrins have been recently implicated in the involvement of pathogenic strain differences in *Drosophila* and in *Culex quinquefasciatus* (a vector for lymphatic filariasis) [86, 87].

Surface-associated molecules are also of interest as potential drug targets. In *w*Bm, 19 putative membrane surface proteins, including the WSP were identified by transmembrane prediction programs. Three of these proteins, not found in *w*Mel, showed a strong similarity to genes found in *Wolbachia* from *Culex*. These three, as well as several other surface proteins, have been cloned, expressed and purified and experiments are underway to further characterize these proteins, including immune reactivity tests with sera from infected animals [Ganatra et al., unpubl.].

A number of other molecules are of interest as potential drug targets. For example, heme produced by biosynthesis from *w*Bm could be vital to worm embryogenesis, as molting and reproduction are controlled by ecdysteroid-like hormones [88], whose synthesis requires heme. Depletion of *Wolbachia* might therefore block molting and/or embryogenesis. As nematodes appear to be unable to synthesize heme, they must obtain it from extraneous sources, such as the media, the food supply, or perhaps via endosymbionts. The heme pathway genes are being cloned and expressed by complementation of *Escherichia coli* deletion mutants [Ganatra et al., unpubl.]. We are also identifying, cloning and expressing the heme transporter molecules as potential drug targets.

The completion of the *w*Bm genome clearly offers suggestions as to which metabolites might be potentially provided by *w*Bm to the nematode and which may be required by the endosymbiont and provided by the nematode. It may be possible to identify drugs already available that might inhibit key biochemical pathways in *Wolbachia,* leading to sterility or killing of the adult worms. Using this information, one can formulate hypotheses about the molecular interaction between the endobacteria and their host. The first published report to make use of the *w*Bm genome in this way used the differential display technique to find nematode genes that were upregulated in response to the depletion of *Wolbachia* [89]. At the time of the study, differential display had an advantage over microarrays in that no prior sequence information is required to discover genes that are differentially regulated in response to some treatment. As the *B. malayi* microarray available at the time had a limited set of genes, the use of arbitrary primers to amplify unknown sequences from control and treated worms was ideal. Based on exclusion methods established in the lab, twelve genes were found to be upregulated in response to the depletion of *Wolbachia* from *L. sigmodontis*. One of the upregulated genes (*Ls-ppe-1*) had homology to the phosphate permease family of proteins which has orthologues in *Caenorhabditis elegans, O. volvulus*, *Acanthocheilonema viteae* and *B. malayi*.

Ls-ppe-1 was shown by qPCR to be upregulated threefold in antibiotic-treated worms over that observed in control (untreated) worms. The upregulation persisted up to a month after removal of tetracycline from the drinking water of treated animals. While the upregulation of *Ls-ppe-1* in female worms showed a bimodal pattern, in male worms the upregulation simply showed an increase in expression beginning between days 3 and 6 of tetracycline treatment. The upregulation was *Wolbachia* dependent as *A. viteae* worms, which are devoid of *Wolbachia*, showed no upregulation of PPE-1 when infected animals were treated with tetracycline. *Ls-ppe-1* was also shown not to be upregulated in response to heat shock or oxidative stress, helping to rule out stress due to dying/dead endobacteria as a cause of the upregulation.

The importance of an upregulation of *Ls-ppe-1* is enhanced when one considers RNA interference (RNAi) results from *C. elegans*. The CeC48A72 phosphate permease (an orthologue of *Ls-ppe-1*) is expressed in adult nematodes and larvae and shows a severe phenotype in RNAi experiments (embryonic lethality, sterility of adult nematodes, and other, undescribed growth and morphological defects) [90] similar to the hallmark phenotypic effects of depleting *Wolbachia* from filarial nematodes [2, 61, 91]. While the *w*Bm genome contains all the genes necessary for the synthesis of nucleotides [72, 92], the endobacteria are sequestered from external sources of phosphate needed for the synthesis of nucleotides by the vesicle membrane synthesized by the worm cell [3, 5, 93–95]. *Ls-ppe-1* might be synthesized by the nematode so that phosphate can be transported into the vesicle containing the *Wolbachia*. When the endobacteria are depleted by antiwolbachial drugs, the homeostasis of nucleotide levels is disturbed and the worm cell attempts to compensate by upregulating the phosphate transporter. It is postulated that the first peak seen in the bimodal upregulation of *Ls-ppe-1* comes from the embryos, which are more sensitive to antiwolbachial treatment [54], with the second peak being the expression from the adult female cells. Understanding how *Ls-ppe-1* and other candidate genes, found by utilizing the information that the *w*Bm genome provides, are involved in the symbiosis between the nematode and *Wolbachia* will be aided by producing functional knock-outs in the nematode by RNAi. RNAi has been successfully established and used to knock down various house-keeping genes and genes important in larval molting in *B. malayi*, *O. volvulus* and *L. sigmodontis* [96–98]. If the *Ls-ppe-1* is essential for *Wolbachia* survival, this will be apparent by histochemistry/electron microscopy as the *Wolbachia* will die if unable to synthesize nucleotides.

In summary, comparative *Wolbachia* genomic sequencing and analysis has, and will continue to provide potential targets for nematode drug discovery and even insect control programs. In addition, the analysis will also provide useful information on the biology and evolution of these and related endosymbionts. Nevertheless, it will be our task to analyze the data, experimentally verify the inferences and effectively utilize the information provided to us.

References

1 World Health Organization: Onchocerciasis and its control. World Health Organ Tech Rep Ser 1995;852:1–10.

2 Molyneux DH, Bradley M, Hoerauf A, Kyelem D, Taylor MJ: Mass drug treatment for lymphatic filariasis and onchocerciasis. Trends Parasitol 2003;19:516–522.

3 McLaren DJ, Worms MJ, Laurence BR, Simpson MG: Micro-organisms in filarial larvae (Nematoda). Trans R Soc Trop Med Hyg 1975;69:509–514.
4 Vincent AL, Portaro JK, Ash LR: A comparison of the body wall ultrastructure of *Brugia pahangi* with that of *Brugia malayi*. J Parasitol 1975;63:567–570.
5 Kozek WJ: Transovarially-transmitted intracellular microorganisms in adult and larval stages of *Brugia malayi*. J Parasitol 1977;63:992–1000.
6 Kozek WJ, Marroquin HF: Intracytoplasmic bacteria in *Onchocerca volvulus*. Am J Trop Med Hyg 1977;26:663–678.
7 Williams SA, Lizotte-Waniewski MR, Foster J, Guiliano D, Daub J, Scott AL, Slatko B, Blaxter M: The filarial genome project: analysis of the nuclear, mitochondrial and endosymbiont genomes of *Brugia malayi*. Int J Parasitol 2000;30:411–419.
8 Sironi M, Bandi C, Sacchi L, Di Sacco B, Damiani G, Genchi, C: Molecular evidence for a close relative of the arthropod endosymbiont *Wolbachia* in a filarial worm. Mol Biochem Parasitol 2000;2:223–227.
9 Bandi C, Trees S, Brattig N: *Wolbachia* in filarial nematodes: evolutionary aspects and implications for the pathogenesis and treatment of filarial disease. Vet Parasitol 2001;98:215–238.
10 Casiraghi M, Bain O, Guerrero R, Martin C, Pocacqua V, Gardner SL Franceschi A, Bandi C: Mapping the presence of *Wolbachia pipientis* on the phylogeny of filarial nematodes: evidence for symbiont loss during evolution. Int J Parasitol 2004;34:191–203.
11 O'Neill SL, Giordano R, Colbert AM, Karr TL, Robertson HM: 16S rRNA phylogenetic analysis of the bacterial endosymbionts associated with cytoplasmic incompatibility in insects. Proc Natl Acad Sci USA 1992;89:2699–2702.
12 Bandi C, Sironi M, Nalepa CA, Corona S, Sacchi: Phylogenetically distant intracellular symbionts in termites. Parassitologia 1997;39:71–75.
13 Werren JH: Biology of *Wolbachia*. Ann Rev Entomol 1997;42:587–609.
14 Vandekerckhove T, Watteyne S, Willems A, Swings JG, Mertens J, Gillis M: Phylogenetic analysis of the 16S rDNA of the cytoplasmic bacterium *Wolbachia* from the novel host *Folsomia candida* (Hexapoda, Collembola) and its implications for wolbachial taxonomy. FEMS Microbiol Lett 1999;180:279–286.
15 Bazzocchi C, Jamnongluk W, O'Neill S, Anderson TJC, Genchi C, Bandi C: *wsp* sequences from the *Wolbachia* of filarial nematodes. Cur Microbiol 2000;4:96–100.
16 Lo N, Casiraghi M, Salati E, Bazzocchi C, Bandi C: How many *Wolbachia* supergroups exist? Mol Biol Evol 2002;3:341–346.
17 Casiraghi M, Bordenstein SR, Baldo L, Lo N, Beninati T, Wernegreen JJ, Werren JH, Bandi C: Phylogeny of *Wolbachia pipientis* based on *glt*A, *gro*EL and *fts*Z gene sequences: clustering of arthropod and nematode symbionts in the F supergroup, and evidence for further diversity in the *Wolbachia* tree. Microbiology 2005;151:4015–4022.
18 Casiraghi M, Anderson TJ, Bandi C, Bazzocchi C, Genchi C: A phylogenetic analysis of filarial nematodes:comparison with the phylogeny of *Wolbachia* endosymbionts. Parasitology 2001;122: 93–103.
19 Plenge-Bonig A, Kromer M, Büttner DW: Light and electron microscopy studies on *Onchocerca jakutensis* and *O. flexuosa* of red deer show different host-parasite interactions. Parasitol Res 1995;81:66–73.
20 Bandi C, Anderson TJ, Genchi C, Blaxter ML: Phylogeny of *Wolbachia* in filarial nematodes. Proc R Soc Lond 1998;265:2407–2413.
21 Taylor MJ, Hoerauf A: *Wolbachia* bacteria of filarial nematodes. Parasitol Today 1999;15: 437–442.
22 Hoerauf A, Nissen-Pahle K, Schmetz C, Henkle-Duhrsen K, Blaxter ML, Büttner DW, Gallin MY, Al-Qaoud KM, Lucius R, Fleischer B: Tetracycline therapy targets intracellular bacteria in the filarial nematode *Litomosoides sigmodontis* and results in filarial infertility. J Clin Invest 1999;103:11–18.
23 Chirgwin SR, Porthousel KH, Nowling JM, Klei TR: The filarial endosymbiont *Wolbachia* sp. is absent from *Setaria equina*. J Parasitol 2002;88:1248–1250.
24 Egyed Z, Sreter T, Szell Z, Nyiro G, Marialigeti K, Varga I: Molecular phylogenetic analysis of *Onchocerca lupi* and its *Wolbachia* endosymbiont. Vet Parasitol 2002;108:153–161.

25 Büttner DW, Wanji S, Bazzocchi C, Bain O, Fischer P: Obligatory symbiotic *Wolbachia* endobacteria are absent from *Loa loa*. Filaria J 2003;2:10.
26 McGarry HF, Pfarr K, Egerton G, Hoerauf A, Akue JP, Enyong P, Wanji S, Klager SL, Bianco AF, Beeching NJ, Taylor MJ: Evidence against *Wolbachia* symbiosis in *Loa loa*. Filaria J 2003;2:9.
27 Brouqui P, Fournier PE, Raoult D: Doxycycline and eradication of microfilaremia in patients with loiasis. Emerg Infect Dis 2001;7(suppl 3):604–605.
28 Grobusch MP, Kombila M, Autenrieth I, Mehlhorn H, Kremsner PG: No evidence of *Wolbachia* endosymbiosis with *Loa loa* and *Mansonella perstans*. Parasitol Res 2003;90:405–408.
29 Taylor MJ, Bilo K, Cross HF, Archer JP Underwood AP: 16S rDNA phylogeny and ultrastructural characterization of *Wolbachia* intracellular bacteria of the filarial nematodes *Brugia malayi*, *B. pahangi*, and *Wuchereria bancrofti*. Exp Parasitol 1999;91:356–361.
30 Taylor MJ, Bandi C, Hoerauf AM, Lazdins J: *Wolbachia* bacteria of filarial nematodes: A target for control? Parasitol Today 2000;16:179–180.
31 Taylor MJ, Cross HF, Bilo K: Inflammatory responses induced by the filarial nematode *Brugia malayi* are mediated by lipopolysaccharide-like activity from endosymbiotic *Wolbachia* bacteria. J Exp Med 2000;191:1429–1436.
32 McGarry HF, Egerton GL, Taylor MJ: Population dynamics of *Wolbachia* bacterial endosymbionts in *Brugia malayi*. Mol Biochem Parasitol 2004;135:57–67.
33 Fenn K, Blaxter M: Quantification of *Wolbachia* bacteria in *Brugia malayi* through the nematode lifecycle. Mol Biochem Parasitol 2004;137:361–364.
34 Jolodar A, Fischer P, Butner DW, Brattig NW: *Wolbachia* endosymbionts of *Onchocerca volvulus* express a putative periplasmic HtrA-type serine protease. Microbes Infect 2004;6:141–149.
35 Bazzocchi C, Ceciliani F, McCall JW, Ricci I, Genchi C, Bandi C: Antigenic role of the endosymbionts of filarial nematodes: IgG response against the *Wolbachia* surface protein in cats infected with *Dirofilaria immitis*. Proc R Soc Lond B Biol Sci 2000;267:2511–2516.
36 Fischer P, Bonow I, Büttner DW, Kamal IH, Liebau E: An aspartate aminotransferase of *Wolbachia* endobacteria from *Onchocerca volvulus* is recognized by IgG1 antibodies from residents of endemic areas. Parasitol Res 2003;90:38–47.
37 Lamb TJ, Le Goff L, Kurniawan A, Guiliano DB, Fenn K, Blaxter ML, Read AF, Allen JE: Most of the response elicited against *Wolbachia* surface protein in filarial nematode infection is due to the infective larval stage. J Infect Dis 2004;189:120–127.
38 Punkosdy GA, Dennis VA, Lasater BL, Tzertzinis G, Foster JM, lammie PJ: Detection of serum IgG antibodies specific for *Wolbachia* surface protein in rhesus monkeys infected with *Brugia malayi*. J Infect Dis 2001;3:385–389.
39 Chirgwin SR, Coleman SU, Porthouse KH, Nowling JM, Punkosdy GA, Klei TR: Removal of *Wolbachia* from *Brugia pahangi* is closely linked to worm death and fecundity but does not result in altered lymphatic lesion formation in Mongolian gerbils (*Meriones unguiculatus*). Infect Immun 2003;71:6986–6994.
40 Francis H, Awadzi K, Ottesen EA: The Mazzotti reaction following treatment of onchocerciasis with diethylcarbamazine: clinical severity as a function of infection intensity. Am J Trop Med Hyg 1985;3:529–536.
41 Ottesen EA: The Wellcome Trust Lecture. Infection and disease in lymphatic filariasis: an immunological perspective. Parasitology 1992;104:S71–S79.
42 Pearlman E: Experimental onchocercal keratitis. Parasitol Today 1996;12:261–267.
43 Freedman D: Immune dynamics in the pathogenesis of human lymphatic filariasis. Parasitol Today 1998;14:229–234.
44 Keiser PB, Reynolds SM, Awadzi K, Ottesen EA, Taylor MJ, Nutman TB: Bacterial endosymbionts of *Onchocerca volvulus* in the pathogenesis of posttreatment reactions. J Infect Dis 2002;6: 805–811.
45 Brattig NW: Pathogenesis and host responses in human onchocerciasis: impact of *Onchocerca* filariae and *Wolbachia* endobacteria. Microbes Infect 2004;6:113–128.
46 Brattig NW, Rathjens U, Ernst M, Geisinger F, Renz A, Tischendorf FW: Lipopolysaccharide-like molecules derived from *Wolbachia* endobacteria of the filaria *Onchocerca volvulus* are candidate mediators in the sequence of inflammatory and antiinflammatory responses of human monocytes. Microbes Infect 2000;2:1147–1157.

47 Saint Andre A, Blackwell NM, Hall LR, Hoerauf A, Brattig NW, Volkmann L, Taylor MJ, Ford L, Hise AG, Lass JH, Diaconu E, Pearlman E: The role of endosymbiotic *Wolbachia* bacteria in the pathogenesis of river blindness. Science 2002;295:1892–1895.
48 Brattig NW, Bazzocchi C, Kirschnin, CJ, Reiling N, Büttner DW, Ceciliani F, Geisinger F, Hochrein H, Ernst M, Wagner H, Bandi C, Hoerauf A: The major surface protein of *Wolbachia* endosymbionts in filarial nematodes elicits immune responses through TLR2 and TLR4. J Immunol 2004;173:437–445.
49 Punkosdy GA, Addiss DG, Lammie PJ: Characterization of antibody responses to *Wolbachia* surface protein in humans with lymphatic filariasis. Infect Immun 2003;71:5104–5114.
50 Pfarr KM, Fischer K, Hoerauf A: Involvement of Toll-like receptor 4 in the embryogenesis of the rodent filaria *Litomosoides sigmodontis*. Med Microbiol Immunol (Berl) 2003;92:53–56.
51 Higazi TB, Filiano A, Katholi CR, Dadzie Y, Remme JH, Unnasch TR: *Wolbachia* endosymbiont levels in severe and mild strains of *Onchocerca volvulus*. Mol Biochem Parasitol 2005;141:109–112.
52 McCall JW, Jun JJ, Bandi C: *Wolbachia* and the antifilarial properties of tetracycline. An untold story. Ital J of Zool 1999;66:7–10.
53 Hoerauf A, Volkmann L, Hamelmann C, Adjei O, Autenrieth IB, Fleischer B, Büttner D: Endosymbiotic bacteria in worms as targets for a novel chemotherapy in filariasis. Lancet 2000;355: 1242–1243.
54 Langworthy NG, Renz A, Mackenstedt U, Henkle-Duhrsen K, de Bronsvoort MB, Tanya VN, Donnelly MJ, Trees AJ: Macrofilaricidal activity of tetracycline against the filarial nematode *Onchocerca ochengi*: elimination of *Wolbachia* precedes worm death and suggests a dependent relationship. Proc Biol Sci 2000;267:1063–1069.
55 Rao R, Weil G: In vitro effects of antibiotics on *Brugia malayi* worm survival and reproduction. J Parasitol 2002;88:605–611.
56 Taylor MJ, Makunde WH, McGarry HF, Turner JD, Mand S, Hoerauf A: Macrofilaricidal activity after doxycycline treatment of *Wuchereria bancrofti*: a double-blind, randomised placebo-controlled trial. Lancet 2005;365:2116–2121.
57 Gilbert J, Nfon CK, Makepeace BL, Njongmeta LM, Hastings IM, Pfarr KM, Renz A, Tanya VN, Trees AJ: Antibiotic chemotherapy of onchocerciasis: in a bovine model, killing of adult parasites requires a sustained depletion of endosymbiotic bacteria (*Wolbachia* species). J Infect Dis 2005;192:1483–1493.
58 Rao RU, Moussa H, Weil GJ: *Brugia malayi*: effects of antibacterial agents on larval viability and development in vitro. Exp Parasitol 2002;101:77–81.
59 Hoerauf A, Mand S, Adjei O, Fleischer B, Büttner DW: Depletion of *Wolbachia* endobacteria in *Onchocerca volvulus* by doxycycline and microfilaridermia after ivermectin treatment. Lancet 2001;357:1415–1416.
60 Hoerauf A, Marfo-Debrekyei Y, Adjei O, Debrah A, Fischer K, Mand S: Loss of worm nests after treatment of *Wuchereria bancrofti* with doxycycline for six weeks suggests a macrofilaricidal effect. Am J Trop Med Hyg 2003;69:S249.
61 Hoerauf A, Mand S, Fischer K, Kruppa T, Marfo-Debrekyei Y, Debrah AY, Pfarr KM, Adjei O, Büttner DW: Doxycycline as a novel strategy against bancroftian filariasis-depletion of Wolbachia endosymbionts from *Wuchereria bancrofti* and stop of microfilaria production. Med Microbiol Immunol (Berl) 2003;192:211–216.
62 Hougard JM, Alley ES, Yameogo L, Dadzie KY, Boatin BA: Eliminating onchocerciasis after 14 years of vector control: a proved strategy. J Infect Dis 2001;184:497–503.
63 Richards FF, Boatin BA, Sauerbrey M: Control of onchocerciasis today; status and challenges. Trends Parasitol 2001;17:558–563.
64 Mackenzie CD, Malecela M, Mueller I, Homeida MA: Approaches to the control and elimination of the clinically important filarial diseases; in Klei TR, Rajan TV (eds): The Filaria. Boston, Kluwer Academic Publishers, 2002, pp 155–165.
65 Ottesen EA: Towards eliminating lymphatic filariasis; in Nutman TB (ed): Lymphatic Filariasis. London, Imperial College Press, 2003, pp 201–205.
66 Dadzie Y, Neira M, Hopkins D: Final report of the conference on the eradicability of onchocerciasis. Filarial J 2003;2:2.
67 Hoerauf A, Büttner DW, Adjei O, Pearlman E: Onchocerciasis. BMJ 2003;326:207–210.

68 Campbell WC: Ivermectin as an antiparasitic agent for use in humans. Annu Rev Microbiol 1991; 45:445–474.
69 Grant W: What is the real target for ivermectin resistance selection in *Onchocerca volvulus*? Parasitol Today 2000;16:458–459; discussion 501–452.
70 Prichard R: Anthelmintic resistance. Vet Parasitol 1994;54:259–268.
71 Prichard R: Genetic variability following selection of *Haemonchus contortus* with anthelmintics. Trends Parasitol 2001;17:445–453.
72 Foster J, Ganatra M, Kamal I, Ware J, Makarova K, Ivanova N, Bhattacharyya A, Kapatral V, Kumar S, Posfai J, Vincze T, Ingram J, Moran L, Lapidus A, Omelchenk M, Kyrpides N, Ghedin E, Wang S, Goltsman E, Joukov V, Ostrovskaya O, Tsukerman K, Mazur M, Comb D, Koonin E, Slatko B: The *Wolbachia* genome of *Brugia malayi*: endosymbiont evolution within a human pathogenic nematode. PLoS Biol 2005;3:e121.
73 Foster JM, Kumar S, Ganatra MB, Kamal IH, Ware J, Ingram J, Pope-Chappell J, Guiliano D, Whitton C, Daub J, Blaxter ML, Slatko BE: Construction of BAC libraries from the parasitic nematode *Brugia malayi* and physical mapping of the genome of its *Wolbachia* endosymbiont. Int J Parasitol 2002;344:733–746.
74 Wu M, Sun LV, Vamathevan J, Riegler M, Deboy R, Brownlie JC, McGraw EA, Martin W, Esser C, Ahmadinejad N, Wiegand C, Madupu R, Beanan MJ, Brinkac LM, Daugherty SC, Durkin AS, Kolonay JF, Nelson WC, Mohamoud Y, Lee P, Berry K, Young MB, Utterback T, Weidman J, Nierman WC, Paulsen IT, Nelson KE, Tettelin H, O'Neill SL, Eisen JA: Phylogenomics of the reproductive parasite *Wolbachia pipientis w*Mel: a streamlined genome overrun by mobile genetic elements. PLoS Biol 2004;2:327–341.
75 GOLD™ Genomes Online Database: http://www.genomesonline.org
76 Sun V, Foster J, Tzertzinis G, Ono M, Bandi C, Slatko B, O'Neill S: Determination of *Wolbachia* genome size by pulsed-field gel electrophoresis. J Bacteriol 2001;183:2219–2225.
77 Liu SL, Hessel A, Sanderson KE: Genomic mapping with I-CeuI, an intron-encoded endonuclease specific for genes for ribosomal RNA in *Salmonella* spp, *Escherichia coli*, and other bacteria. Proc Natl Acad Sci USA 1993;90:6874–6878.
78 http://www.sanger.ac.uk/Projects/Wolbachia
79 Salzberg SL, Dunning Hotopp JC, Delcher AL, Pop M, Smith DR, Eisen MB, Nelson WC: Serendipitous discovery of *Wolbachia* genomes in multiple *Drosophila* species. Genome Biol 2005;6:R23.
80 Salzberg SL, Dunning Hotopp JC, Delcher AL, Pop M, Smith DR, Eisen MB, Nelson WC: Correction: Serendipitous discovery of *Wolbachia* genomes in multiple *Drosophila* species. Genome Biol 2005;6:402.
81 Iturbe-Ormaexte I, Riegler M, O'Neill SL: New names for old strains? *Wolbachia w*Sim is actually *w*Ri. Genome Biol 2005;6:401.
82 Ghedin E, Wang S, Foster J, Slatko B: First sequenced genome of a parasitic nematode. Trends Parasitol 2004;20:151–153.
83 http://www.sanger.ac.uk/Projects/W_pipientis
84 Mavingui P, Van VT, Labeyrie E, Rances E, Vavre F, Simonet P: Efficient procedure for purification of obligate intracellular *Wolbachia pipientis* and representative amplification of its genome by multiple-displacement amplification. App Environ Microbiol 2005;71:6910–6917.
85 Rajan TV: Exogenous nucleosides are required for the morphogenesis of the human filarial parasite *Brugia malayi*. J Parasitol 2004;90:1184–1185.
86 Sinkins S, Walker T, Lynd A, Makepeace B, Godfray H, Parkhill J: *Wolbachia* variability and host effects on crossing type in Culex mosquitoes. Nature 2005;436:257–260.
87 Iturbe-Ormaexte I, Burke G, Riegler M, O'Neill S: Distribution, expression, and motif variability of ankyrin domain genes in *Wolbachia pipientis*. J Bacteriology 2005;187:5136–5145.
88 Warbrick EV, Barker GC, Rees HH, Howells RE: The effect of invertebrate hormones and potential hormone inhibitors on the third larval moult of the filarial nematode, *Dirofilaria immitis*, in vitro. Parasitology 1993;107:459–463.
89 Heider U, Blaxter ML, Hoerauf A, Pfarr KM: Differential display of genes expressed in the filarial nematode *Litomosoides sigmodontis* reveals a putative phosphate permease up-regulated after depletion of *Wolbachia* endobacteria. Int J Med Microbiol 2006;296:287–299.

90 Maeda I, Kohara Y, Yamamoto M, Sugimoto A: Large-scale analysis of gene function in *Caenorhabditis elegans* by high-throughput RNAi. Curr Biol 2001;11:171–176.
91 Hoerauf A, Mand S, Volkmann L, Büttner M, Marfo-Debrekyei Y, Taylor M, Adjei O, Büttner DW: Doxycycline in the treatment of human onchocerciasis: kinetics of *Wolbachia* endobacteria reduction and of inhibition of embryogenesis in female *Onchocerca* worms. Microbes Infect 2003;5: 261–273.
92 Pfarr K, Hoerauf A: The annotated genome of *Wolbachia* from the filarial nematode *Brugia malayi*: What it means for progress in antifilarial medicine. PLoS Med 2005;2:e110.
93 Kozek WJ, Figueroa Marroquin H: Intracytoplasmic bacteria in Onchocerca volvulus. Am J Trop Med Hyg 1977;26:663–678.
94 McLaren DJ: Ultrastructural studies on microfilariae (Nematoda: Filarioidea). Parasitol 1972;65: 317–332.
95 Franz M, Büttner DW: Histology of adult *Brugia malayi*. Trop Med Parasitol 1986;37:282–285.
96 Aboobaker AA, Blaxter ML: Use of RNA interference to investigate gene function in the human filarial nematode *Brugia malayi*. Mol Biochem Parasitol 2003;128:41–51.
97 Lustigman S, Zhang J, Liu J, Oksov Y, Hashmi S: RNA interference targeting cathepsin L and Z-like cysteine proteases of *Onchocerca volvulus* confirmed their essential function during L3 molting. Mol Biochem Parasitol 2004;138:165–170.
98 Pfarr K, Heider U, Hoerauf A: RNAi mediated silencing of actin expression in adult *Litomosoides sigmodontis* is specific, persistent and results in a phenotype. Int J Parasitol 2006;36:661–669.

Kenneth Pfarr
Institute for Medical Microbiology, Immunology and Parasitology
University Clinic Bonn, Sigmund-Freud-Strasse 25
DE–53105 Bonn (Germany)
Tel. +49 228 287 11510, Fax +49 228 287 14330, E-Mail pfarr@parasit.meb.uni-bonn.de

Hoerauf A, Rao RU (eds): Wolbachia.
Issues Infect Dis. Basel, Karger, 2007, vol 5, pp 66–76

Coexist, Cooperate and Thrive: *Wolbachia* as Long-Term Symbionts of Filarial Nematodes

Katelyn Fenn, Mark Blaxter

Institutes of Immunology and Infection Research and Evolutionary Biology, Ashworth Laboratories, University of Edinburgh, Edinburgh, UK

Abstract

Wolbachia of arthropods and nematodes have contrasting patterns of association with their hosts. These patterns reflect the biological nature of the associations: parasitic in arthropods and possibly mutualistic in filarial nematodes. Genome sequence data are aiding in both the resolution of the evolutionary history of *Wolbachia* strains, and in revealing the possible physiological bases of the interactions between bacterium and host. In the filarial nematodes there is evidence for long and stable association between nematode lineages and *Wolbachia* lineages. This long association is reflected in the presence in the host nematode genome of fragments of *Wolbachia* genes horizontally transferred. These genes are functionally inactive due to insertions, deletions and substitutions. For transferred genes to be functional in the host genome, many modifications would need to occur to ensure transcription and correct processing. It is unlikely that such gene transfers would result in the replacement of the *Wolbachia* gene product requirement and so gene transfer can not explain the apparent loss of *Wolbachia* in certain lineages.

The parasitic mode of life has evolved many times. Parasitism is one part of a complex spectrum of interactions between organisms, from predation/herbivory to mutualistic symbiosis. A parasite (or pathogen) is usually defined as a smaller organism that exploits a larger one, the host, for its benefit. While the parasite may share some life goals with its host (Richard Dawkin's 'desiderata' [1, 2]), there is an essential conflict over investment in reproduction: a parasite will promote its reproduction at the expense of the host (and vice versa).

The evolutionary antecedents of parasitism remain unclear. One can imagine casual interactions becoming integrated into a parasite's life history as its reproductive success is promoted by exploitation of the host. In this model, originally free-living organisms evolve to become parasites. Evolution in the opposite direction, from parasitism to free-living habits, is thought to be less possible, as parasites often lose genes for core requirements now supplied by the host. Another possible route away from parasitism is to evolve towards mutualism. Long coexistence between parasite and host can result in the evolution of lowered virulence, and thus of benign or commensal organisms, but the parasitic mode of life remains the basis of the interaction. However, if the parasite provides some adaptive metabolic or other services to the host, the two organisms may have enough shared desiderata for the symbiosis to lose features of pathogenesis and virulence as the parasite is 'tamed' or enslaved.

There are many examples of such mutualistic interactions between metazoans and bacteria, such as the *Buchnera* – aphid interaction, where the bacterium is hosted because of its ability to supply essential amino acids to the otherwise compromised insect [3]. In return, the aphid ensures the transmission of *Buchnera* to its offspring, provides a specialised cellular niche for the bacteria (the bacteriosome) and supplies the bacteria with essential nutrients. 'Curing' the aphid of *Buchnera* with antibiotic is fatal to the insect [4].

Wolbachia of arthropods are characterised by maternal-offspring transmission through the oocyte, and, by various reproductive manipulations, the *Wolbachia* promote the relative fitness of infected female hosts in a mixed population of carriers and non-carriers [5]. Thus the *Wolbachia* 'persuades' its host that, in the short term, they share desiderata, of generation of as many (infected) offspring as is possible, and that this is best achieved by promotion of infection per se. In interactions between *Wolbachia* and arthropods, antibiotic cure (usually with tetracycline) is usually successful, and reveals the otherwise hidden cost to the host of maintenance of the bacterial parasite. In rare cases, possibly due to a host's over-compensation for the effects of the *Wolbachia*, cure results in sterility. In these cases, then, the *Wolbachia* behave as essential genetic elements, and the symbiosis is effectively mutualist, though it may be supposed that it is being driven by the *Wolbachia*. The results of this arms race between *Wolbachia* and its many arthropod hosts can be seen in the evidence for extensive selective sweeps of cytoplasmic genes in arthropod populations and species [6], the variety of parasitic manipulations performed by the *Wolbachia*, and the evolution of resistance in some hosts.

Wolbachia as a Mutualistic Partner

In the nematode *Wolbachia*, the patterns of host infection are very different. *Wolbachia* have been found only in parasitic filarial nematodes of the

Onchocercidae, a family that includes several major human pathogens such as *Onchocerca volvulus* (causing river blindness) and *Wuchereria bancrofti* (lymphatic filariasis and elephantiasis) [7, 8]. Almost all the species within the Onchocercidae harbour *Wolbachia*, and within infected species all individuals carry the symbiont. The phylogenetic relationships of the nematode *Wolbachia* and their nematode hosts are congruent, and thus closely related filarial nematodes have closely related *Wolbachia* [7, 8]. This congruent pattern suggests a long-term, stable relationship between host and symbiont where transfer of the symbiont is exclusively through vertical transmission.

Such long-term co-existence is more likely to lead to or be a feature of mutualism between organisms [9]. In support of a mutualistic association, tetracycline-curing experiments in filarial nematodes have a very different outcome to that observed in arthropod *Wolbachia* infections. Tetracycline treatment of filarial nematodes leads to the death of bacteria, and in abnormalities in nematode growth, moulting, fecundity and lifespan [10]. That these abnormalities are due to the loss of the *Wolbachia* is supported by treatment of related filarial nematode species that lack *Wolbachia*, where no effects of antibiotic treatment are observed [10]. This potentially obligate association is being taken advantage of in the development of novel filariasis chemotherapy regimes, where doxycycline is added to the standard set of anti-filarial drugs. Very positive results have been obtained in trials in infected communities [11–13]. This success raises the obvious question as to whether there are other targets in *Wolbachia* that could be used to treat and cure human filarial disease. The sequence of the complete genome of *w*Bm, the *Wolbachia* from the human-parasitic filarial nematode *Brugia malayi* has been determined [14], as has the genome sequence of *w*Mel from *Drosophila melanogaster* [15]. Other *Wolbachia* genome sequencing projects are also underway [16]. These genomes facilitate wholistic analysis of the evolution and metabolic capacities of the endosymbionts.

On the Origins of *Wolbachia*

Wolbachia are members of the Anaplasmataceae in the Rickettsiales. Related genera are all intracellular pathogens and symbionts (of unknown or parasitic habits) of other eukaryotes, including mammals and arthropods. *Wolbachia* have been identified, using molecular data (mainly the *ftsZ* and 16S ribosomal RNA genes), from diverse hosts, and eight major subclades (named A to H) defined. A to D are considered the major supergroups with A and B *Wolbachia* found only in arthropods (mainly Insecta) and the C and D *Wolbachia* found only in onchocercid nematodes. Clade E was described from springtails (Collembola), G from Australian spiders and H from termites. Clade F Wolbachia occur in the

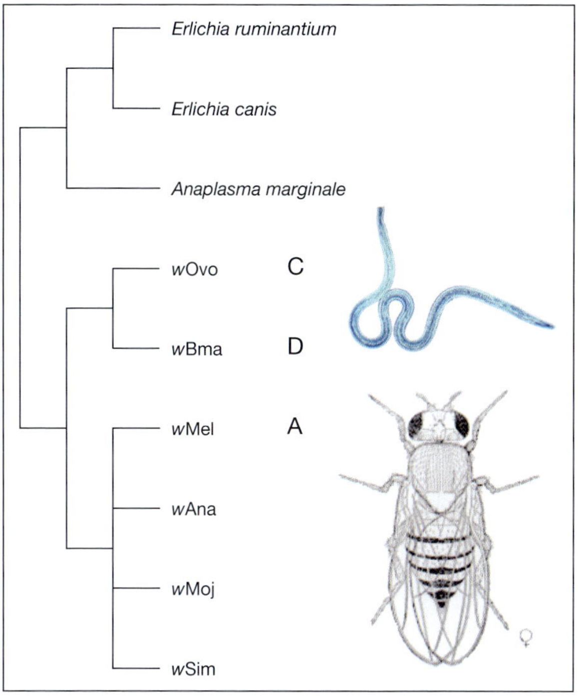

Fig. 1. The relationships of the *Wolbachia* of arthropods to those from nematodes. Using sequences from 43 protein-coding and ribosomal RNA genes, the relationships of arthropod *Wolbachia* (the clade A strains *w*Mel, *w*Sim, *w*Moj and *w*Ana) were compared to those from nematodes (*w*Ovo from clade C and *w*Bm from clade D). The *Anaplasma* and *Ehrlichia* species are outgroups [19].

widest range of host types including termites, a weevil, cimicid bugs, North American bush crickets and onchocercid nematodes (*Mansonella* spp.) [17, 18].

Understanding the relationships between *Wolbachia* will answer some fundamental questions about the biology of *Wolbachia,* in particular the pattern of change of the parasitic versus mutualistic phenotype, in particular whether the mutualist nematode strains derive from parasitic arthropod-infecting ancestors, as has usually been assumed. As phylogenetic analysis using a small number of genes has not been able to robustly root the tree [8], we used partial genome sequence available from *w*Ovo, a C clade *Wolbachia* from *O. volvulus*, to identify 42 orthologous genes in *w*Ovo, *w*Mel, *w*Bm and related anaplasmataceae, and subjected this dataset to phylogenetic analysis [19]. The results suggest that the ancestor of all extant *Wolbachia* was probably an intracellular parasite, because the root robustly falls between the arthropod clade A and the nematode clades C and D (fig. 1).

Thus, nematode *Wolbachia* share a last common ancestor (presumably also a nematode endosymbiont) before they share a common ancestor with the major A and B clade arthropod parasites. This suggests that the mutualistic features of the nematode-*Wolbachia* interaction have evolved from parasitic antecedents.

We await with much interest the acquisition of genome sequence data from additional *Wolbachia* clades, particularly E, F, G and H, and these rarer strains may assist in resolving the early history of *Wolbachia*. It is particularly intriguing that clade F hosts include both arthropods and onchocercid nematodes. Based on the lack of recombination observed in this clade [18], it has been suggested that clade F *Wolbachia* are mutualistic with their hosts, as a mutualistic lifestyle reduces the likelihood of horizontal transfer events between hosts and so limits the opportunity for recombination. It is striking that the nematode infected with a clade F *Wolbachia* is an onchocercid. The possibility that *Wolbachia* have been horizontally transferred between nematode and arthropod recently within the F clade needs to be investigated: perhaps an arthropod *Wolbachia* has invaded and replaced the original clade C *Wolbachia* that would be expected in *Mansonella*?

Gene Loss and Metabolic Dependency in Long Associations: Why Do *Wolbachia* Need Nematodes?

As a result of long-term co-evolution, intracellular endosymbionts often lose genes and become dependent on their host cell for many basic metabolic processes [20]. Gene loss in symbionts is piecemeal, and proceeds through inactivation through point mutation (substitutions and small insertions or deletions) followed by progressive loss of the inactive gene's DNA. Long-term stable endosymbiotic relationships are thus likely to result in reduced genomes that encode compromised inferred metabolism. In the process that leads to this end-state, the symbiont genomes may contain many degenerating pseudogenes. Another common finding in symbiont genomes is that the proportion of repeats is also reduced, and that prophages, insertion sequences and other mobile genetic elements are eliminated [21].

The *w*Bm genome is smaller that that of *w*Mel, and encodes fewer protein-coding genes. Genome degradation is ongoing in *w*Bm: it has 98 predicted pseudogenes. It is often easy to work out why a bacteria needs to live inside a cell just by looking at genome content. For example, *w*Bm does not have the capacity to synthesise all of the amino acids required for protein synthesis and so must import these from the host cell. Other metabolic dependencies can be identified, and may go towards the development of an extracellular culture system for *Wolbachia*, which would be a very useful research tool.

A surprising finding in both sequenced *Wolbachia* genomes is that they contain a large proportion of repeated sequences. In *w*Mel these are associated with insertion elements, and make up 14% of the genome [15]. Although this association with insertion elements is not apparent in *w*Bm, repeated sequences still make up 5.4% of the genome [14]. While uncommon in bacteria, this large proportion of repeat sequences is seen in other members of the Rickettsiales and so may be a feature of this family. These repeats may aid in gene duplication and recombination, which in turn may lead to novelty and plasticity in the genome of the bacteria. Alternatively, the loss of some elements of DNA repair metabolism may have permitted these elements to proliferate, possibly to the detriment of the fitness of the *Wolbachia*.

Why Do Filarial Nematodes Need *Wolbachia*: Vitamins and Smokescreens?

The apparently essential role for *Wolbachia* revealed by treatment of filarial nematodes with tetracycline remains enigmatic. It cannot be simply supply of an otherwise unavailable metabolic service, as the existence of filarial species that lack *Wolbachia* (*Loa loa* and *Onchocerca flexuosa*, for example) shows that successful filarial parasites can survive without endosymbiotic support. Comparisons of the metabolic capacity of *w*Bm with that of its host (the complete genome of *B. malayi* is being sequenced) offers some suggestive, though not compelling, possible roles for *w*Bm in assisting its host [14, 16]. *B. malayi* are apparently unable to synthesize riboflavins or haem endogenously, but *w*Bm has a full set of genes for riboflavin and haem synthesis and so may provide these important co-factors. Haem is an essential vitamin for other nematodes in culture (for example *Caenorhabditis elegans* [22]) and the *Wolbachia* may be the filarial nematodes' way of ensuring supply of this limiting nutrient in the mammalian or arthropod host. A link with the effects of tetracycline treatment on moulting may be through the requirement for haem in cytochromes involved in the production of steroid moulting hormones. Unlike other bacterial endosymbionts, such as some rickettsia and *Buchnera aphidicola*, the *w*Bm genome contains complete pathways for de novo biosynthesis of purines and pyrimidines. The *Wolbachia* may produce nucleotides for both internal consumption and also to supplement the nucleotide pool of the nematode host during oogenesis and embryogenesis. Again, tetracycline treatment has significant effects on fecundity.

Filarial nematodes induce a profound, specific immuno-unresponsiveness in their mammalian hosts, and the *Wolbachia* may play a part in inducing or maintaining this state. Early mammalian responses to *Wolbachia*-infected filarial nematodes include responses to bacterial components, and it has been suggested that the *Wolbachia* may be skewing the mammalian response to ineffective anti-bacterial pathways, and protecting the nematode. Candidate molecules that may be involved in the deception of the immune system are yet to be identified unequivocally, but *Wolbachia* surface protein and surface glycolipoproteins/glycolipids are strong candidates [16, 23–25].

Cementing the *Wolbachia*-Nematode Relationship with Molecular Glue

How do *Wolbachia* interact with, and manipulate their hosts? By comparing the genomes of parasitic *w*Mel and mutualistic *w*Bm, groups of genes hypothesized to be involved in host interactions have been identified. A type IV secretion system may have a role in translocating *Wolbachia* gene products to the host cell environment, as type IV secretion systems in other bacteria export virulence factors. The components secreted by the *Wolbachia* type IV systems are as yet unknown, but may be found amongst the genes predicted to encode secreted proteins, genes closely linked to the type IV operon, and genes located within prophages. Candidate effector molecules include a family of ankyrin-repeat-containing proteins (ANK proteins). The ankyrin repeat mediates protein-protein interactions, and ANK proteins have been implicated in host-pathogen interactions in other systems. *w*Mel contains an unusually high number of ANK proteins (over 25) [15, 16] and these proteins may be directly involved in cell cycle manipulations through interaction with host cell regulators that also contain ankyrin repeats. *w*Bm only contains nine ANK proteins [14]. Interestingly, in the arthropod *Wolbachia*, some ANK proteins are associated with active phage, and thus may be transferred between bacterial genomes. In a mosquito model system, cytoplasmic incompatibility phenotypes segregate with distinct ANK protein types [26].

Are Filarial *Wolbachia* Really Obligate Mutualists?

Although many nematode-bacterial symbioses exist, *Wolbachia* infection seems to be exclusive to the Onchocercidae. There is conflicting evidence for the mutualist status of this symbiosis. While there is general congruence between molecular phylogenies of filarial nematodes and their *Wolbachia*, some filarial species that lack symbionts are nested within symbiont-containing clades [7, 8]. Mapping these events onto the nematode phylogeny, and assuming a single event of initial infection of an onchocercid ancestor implies that there have been six independent events of loss of *Wolbachia*. For example, *O. volvulus* and most other *Onchocerca* species carry *Wolbachia* but *O. flexuosa*, a red deer parasite, appears to lack *Wolbachia* altogether [27]. This pattern would be unexpected for an essential association, as genes and functions lost by either partner cannot easily be regained. If the nematodes can adapt to *Wolbachia* loss, then the relationship is not truly essential, and treatment regimes in infected human communities may generate *Wolbachia*-free nematode populations. Until the relationship is fully understood it is difficult to foresee what antibacterials may do to the pattern of *Wolbachia* currently seen in filarial nematodes.

Evidence of Absence: Horizontal Gene Exchange between *Wolbachia* and Nematode Genomes

Symbionts can leave their marks on the genomes of their hosts in different ways. Hosts may lose genetic material as they come to rely upon services provided by the symbiont. This will tend to fix the relationship. Alternatively, the host could acquire genes from the symbiont, and thus evolve an autochthonous source of the functions that the symbiont provides. While horizontal gene transfer from bacteria to eukaryotes has been demonstrated, and demonstrated particularly compellingly in nematodes parasitic on plants [28], it is rare in general. To move a genenetic function from the symbiont chromosome to the host nuclear genome requires DNA transfer, integration, and evolution of eukaryotic gene expression signals: eukaryotic promoter sequences appropriately placed, transcription start sites following an eukaryotic model, the emplacement of introns to promote pre-mRNA processing and stability, and acquisition of polyadenylation signals. The first two steps (transfer and integration) are not unlikely, as there is ample evidence of incorporation of (non-functional) fragments of the metazoan mitochondrial genome into nuclear sites in many species [29]. The cytoplasmic location of *Wolbachia* suggests that its DNA may similarly be taken up and integrated. This has now been observed twice, once in an adzuki bean beetle [30], where a large fragment of an A-clade *Wolbachia* genome is located on the host sex chromosome, and now also in *O. volvulus*, where a small fragment of *Wolbachia* sequence has been detected, lying upstream of a TATA-box binding protein gene [19]. This insertion is not detectable in *B. malayi* (fig. 2), but is found in *Onchocerca ochengi*, suggesting that the horizontal gene transfer occurred in an *Onchocerca* ancestor. If it was inserted before the last common ancestor of *O. flexuosa*, *O. volvulus* and *O. ochengi*, its presence would be prima facie evidence that *O. flexuosa* once had, and has now lost, *Wolbachia* symbionts. The fragment is a relict, made non-functional

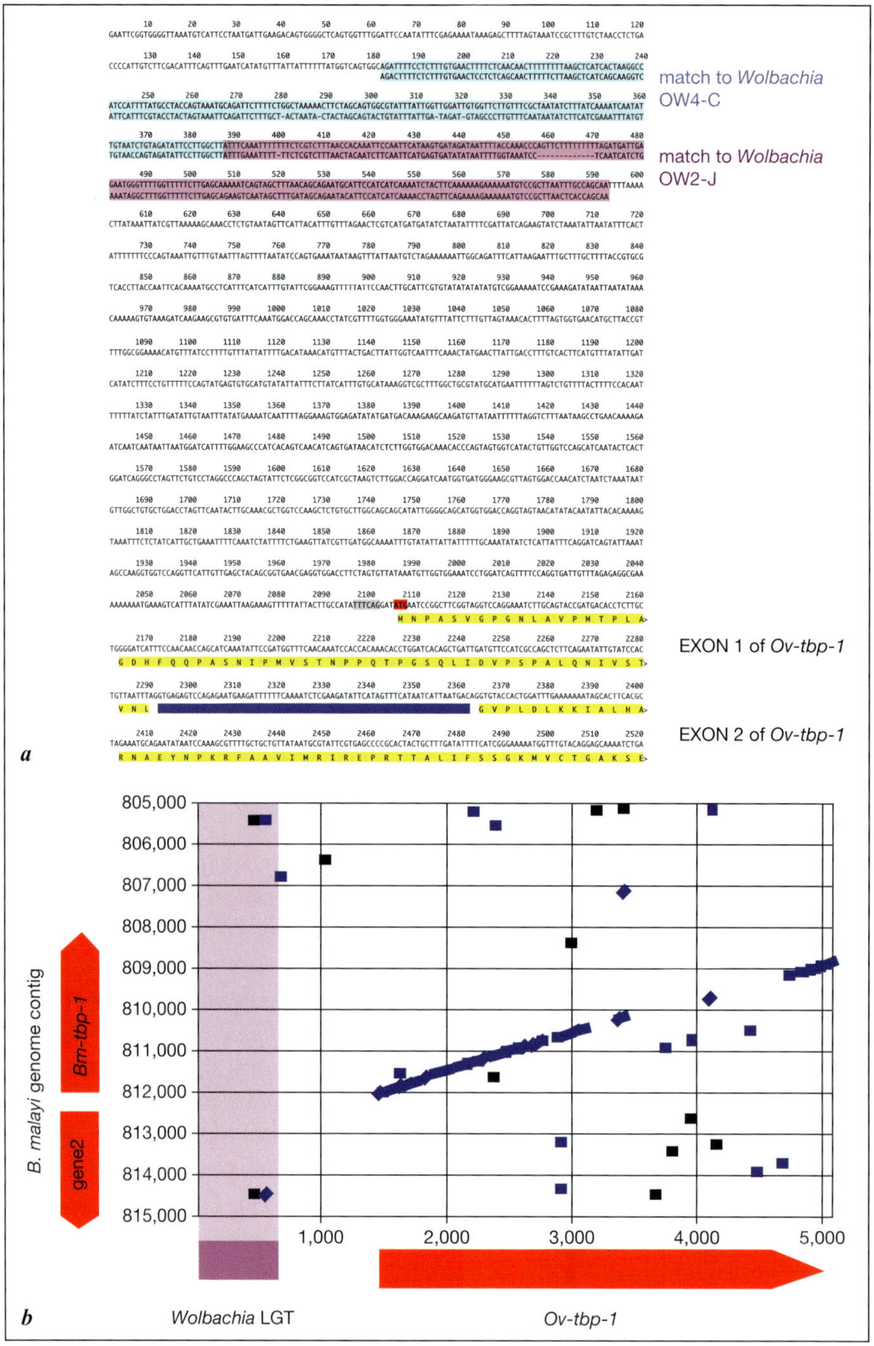

match to Wolbachia OW4-C
match to Wolbachia OW2-J
EXON 1 of Ov-tbp-1
EXON 2 of Ov-tbp-1
a
805,000
806,000
807,000
808,000
809,000
810,000
811,000
812,000
813,000
814,000
815,000
1,000
2,000
3,000
4,000
5,000
B. malayi genome contig
Bm-tbp-1
gene2
b
Wolbachia LGT
Ov-tbp-1

Fig. 2. A *Wolbachia* genomic insertion in the genome of *O. volvulus* is absent from *B. malayi*. ***a*** The sequence of the region of the *O. volvulus* nuclear genome upstream of the TATA-binding protein gene (*Ov-tbp-1*; the *trans*-splice acceptor site is highlighted in grey, the initiation methionine codon in red, the translation of the two exons in yellow, and the intron in dark blue) is compared to two fragments of the *Wolbachia* *w*Ovo genome (gene OW4-C, of unknown function, highlighted in light blue, and gene OW2-J, encoding a phosphomannomutase, in purple). In the first part of the figure, the lower sequence derives from *w*Ovo [19]. ***b*** The *Wolbachia* genome insertion is missing from the orthologous region of the *B. malayi* genome. In this dot-plot, each dot indicates a region ≥20 bp with ≥60% identical residues. Blue dots indicate matches in the same orientation, while black dots indicate matches in the reverse orientation. The red arrows indicate the positions of the protein-coding genes on the *O. volvulus* and *B. malayi* nuclear DNA fragments, and the mauve highlighting indicates the position of the *Wolbachia* lateral gene transfer event (LGT) in *O. volvulus*. While the exons of the TATA-binding protein gene are highly conserved between the two filarial nematodes, the *Wolbachia* LGT is absent from *B. malayi*. Another gene ('gene2') is found upstream of the TATA-binding protein gene in *B. malayi*.

by insertions, deletions and substitutions, of two partial genes, and is not predicted to encode any functional protein. We regard it extremely unlikely that *Wolbachia* genes encoding mutualism functions will have been transferred intact to the host nucleus and thus substitute for the presence of *Wolbachia* in nematodes that appear to have lost their symbionts.

References

1 Dawkins R: The Extended Phenotype. Oxford, Oxford University Press, 1982.
2 Dawkins R: Parasites, desiderata lists and the paradox of the organism. Parasitology 1990;100(suppl):S63–S73.
3 Shigenobu S, Watanabe H, Hattori M, Sakaki Y, Ishikawa H: Genome sequence of the endocellular bacterial symbiont of aphids *Buchnera* sp. APS. Nature 2000;407:81–86.
4 Baumann P, Baumann L, Clark MA, Thao ML: *Buchnera aphidicola*: the endosymbiont of aphids. ASM News 1998;64:203–209.
5 Werren JH: Biology of *Wolbachia*. Annu Rev Entomol 1997;42:587–609.
6 Jiggins FM: Male-killing *Wolbachia* and mitochondrial DNA: selective sweeps, hybrid introgression and parasite population dynamics. Genetics 2003;164:5–12.
7 Casiraghi M, Anderson TJ, Bandi C, Bazzocchi C, Genchi C: A phylogenetic analysis of filarial nematodes: comparison with the phylogeny of *Wolbachia* endosymbionts. Parasitology 2001;122: 93–103.
8 Casiraghi M, Bordenstein SR, Baldo L, Lo N, Beninati T, et al: Phylogeny of *Wolbachia pipientis* based on gltA, groEL and ftsZ gene sequences: clustering of arthropod and nematode symbionts in the F supergroup, and evidence for further diversity in the *Wolbachia* tree. Microbiology 2005;151:4015–4022.
9 Fenn K, Blaxter M: Are filarial nematode *Wolbachia* obligate mutualist symbionts? Trends Ecol Evol 2004;19:163–166.
10 Hoerauf A, Nissen-Pahle K, Schmetz C, Henkle-Duhrsen K, Blaxter ML, et al: Tetracycline therapy targets intracellular bacteria in the filarial nematode *Litomosoides sigmodontis* and results in filarial infertility. J Clin Invest 1999;103:11–18.

11 Hoerauf A, Mand S, Adjei O, Fleischer B, Buttner DW: Depletion of *Wolbachia* endobacteria in *Onchocerca volvulus* by doxycycline and microfilaridermia after ivermectin treatment. Lancet 2001;357:1415–1416.
12 Taylor MJ, Makunde WH, McGarry HF, Turner JD, Mand S, et al: Macrofilaricidal activity after doxycycline treatment of *Wuchereria bancrofti*: a double-blind, randomised placebo-controlled trial. Lancet 2005;365:2116–2121.
13 Turner JD, Mand S, Debrah AY, Muehlfeld J, Pfarr K, et al: A randomized, double-blind clinical trial of a 3-week course of doxycycline plus albendazole and ivermectin for the treatment of *Wuchereria bancrofti* infection. Clin Infect Dis 2006;42:1081–1089.
14 Foster J, Ganatra M, Kamal I, Ware J, Makarova K, et al: The *Wolbachia* genome of *Brugia malayi*: endosymbiont evolution within a human pathogenic nematode. PLoS Biol 2005;3:e121.
15 Wu M, Sun LV, Vamathevan J, Riegler M, Deboy R et al: Phylogenomics of the reproductive parasite *Wolbachia pipientis w*Mel: a streamlined genome overrun by mobile genetic elements. PLoS Biol 2004;2:E69.
16 Fenn K, Blaxter M: *Wolbachia* genomes: revealing the biology of parasitism and mutualism. Trends Parasitol 2006;22:60–65.
17 Lo N, Casiraghi M, Salati E, Bazzocchi C, Bandi C: How many *Wolbachia* Supergroups Exist? Mol Biol Evol 2002;19:341–346.
18 Panaram K, Marshall JL: F supergroup *Wolbachia* in bush crickets: what do patterns of sequence variation reveal about this supergroup and horizontal transfer between nematodes and arthropods? Genetica 2006, Epub ahead of print.
19 Fenn K, Conlon C, Jones M, Quail MA, Holroyd NE, et al: Phylogenetic relationships of the *Wolbachia* of nematodes and arthropods. PLoS Pathog 2006, Epub ahead of print.
20 Tamas I, Klasson LM, Sandstrom JP, Andersson SG: Mutualists and parasites: how to paint yourself into a (metabolic) corner. FEBS Lett 2001;498:135–139.
21 Frank AC, Amiri H, Andersson SG: Genome deterioration: loss of repeated sequences and accumulation of junk DNA. Genetica 2002;115:1–12.
22 Vanfleteren JR: Nematode growth factor. Nature 1974;248:255–257.
23 Taylor MJ, Cross HF, Bilo K: Inflammatory responses induced by the filarial nematode *Brugia malayi* are mediated by lipopolysaccharide-like activity from endosymbiotic *Wolbachia* bacteria. J Exp Med 2000;191:1429–1436.
24 Taylor MJ, Cross HF, Ford L, Makunde WH, Prasad GB, et al: *Wolbachia* bacteria in filarial immunity and disease. Parasite Immunol 2001;23:401–409.
25 Brattig NW, Bazzocchi C, Kirschning CJ, Reiling N, Buttner DW, et al: The major surface protein of *Wolbachia* endosymbionts in filarial nematodes elicits immune responses through TLR2 and TLR4. J Immunol 2004;173:437–445.
26 Sinkins SP, Walker T, Lynd AR, Steven AR, Makepeace BL, et al: *Wolbachia* variability and host effects on crossing type in *Culex mosquitoes*. Nature 2005;436:257–260.
27 Brattig NW, Buttner DW, Hoerauf A: Neutrophil accumulation around Onchocerca worms and chemotaxis of neutrophils are dependent on *Wolbachia* endobacteria. Microbes Infect 2001;3: 439–446.
28 Scholl EH, Thorne JL, McCarter JP, Bird DM: Horizontally transferred genes in plant-parasitic nematodes: a high-throughput genomic approach. Genome Biol 2003;4:R39.
29 Richly E, Leister D: NUMTs in sequenced eukaryotic genomes. Mol Biol Evol 2004;21:1081–1084.
30 Kondo N, Nikoh N, Ijichi N, Shimada M, Fukatsu T: Genome fragment of *Wolbachia* endosymbiont transferred to X chromosome of host insect. Proc Natl Acad Sci USA 2002;99:14280–14285.

Katelyn Fenn
Institutes of Immunology and Infection Research and Evolutionary Biology
Ashworth Laboratories, University of Edinburgh
Edinburgh EH9 3JT (UK)
Tel. +44 131 651 3618, Fax +44 131 650 6564, E-Mail k.fenn@ed.ac.uk

Hoerauf A, Rao RU (eds): Wolbachia.
Issues Infect Dis. Basel, Karger, 2007, vol 5, pp 77–89

Insights into *Wolbachia* Biology Provided through Genomic Analysis

R. Yamada, J.C. Brownlie, E.A. McGraw, S.L. O'Neill

School of Integrative Biology, The University of Queensland, Brisbane, Australia

Abstract

Wolbachia are maternally inherited intracellular bacteria that infect a range of invertebrates, including insects, mites, spiders and nematodes. They influence the biology of their host through a range of different mechanisms, from nutritional mutualism to various forms of reproductive parasitism. The recent partial and complete sequencing of a number of *Wolbachia* genomes is providing a wealth of comparative data that can be used to better understand the biology of these organisms, from providing putative genes and mechanisms involved in host interaction through to new polymorphic markers with which to better understand *Wolbachia* ecology.

Over the last 6 years, several insect bacterial endosymbiont genomes have been fully sequenced, including *Wolbachia* genomes from the insect *Drosophila melanogaster* (*w*Mel) and the filarial nematode, *Brugia malayi* (*w*Bm) [1, 2]. In addition, partial genome sequences of two *Wolbachia* strains *w*Ana and *w*Ri that infect different *Drosophila* host species have been assembled as by-products of host insect genome sequencing projects [3, 4]. This unexpected discovery suggests that additional *Wolbachia* genomes will likely be sequenced in the course of future insect genome projects. At the present time, eight dedicated genome sequencing projects of different *Wolbachia* strains are ongoing that aim to sequence *Wolbachia* from a range of hosts including *Drosophila* (*w*Ana, *w*No, *w*Ri), *Muscidifurax uniraptor* (*w*Uni), *Armadillidium vulgare* (*w*Vul), *Culex pipiens* (*w*Pip), *Dirofilaria immitis* (*w*Di), and *Onchocerca volvulus* (*w*Ov). This surge in *Wolbachia* genome data has found a strong framework for comparative studies in the completed sequencing projects of closely related pathogenic *Rickettsia* [5–7].

Although two whole genome sequences of *Wolbachia* have been completed to date, the progress for other *Wolbachia* sequencing projects is slow due to difficulties in obtaining large amounts of pure *Wolbachia* genomic DNA. For genome sequencing, several strategies to obtain genomic DNA have been employed including pulsed-field gel electrophoresis, subtractive hybridization to isolate cloned *Wolbachia* DNA from whole host genomic libraries or long PCR-based approaches using PCR primers designed from the *w*Mel genome [1, 2]. These methodologies are required due to the inability to culture *Wolbachia* on cell-free media. Recently, an efficient strategy for the isolation of *Wolbachia* genomic DNA has been reported. Through a combination of differential centrifugation, pulsed-field gel electrophoresis, and whole genome amplification by multiple-displacement amplification, large quantities of *Wolbachia* DNA, suitable for genome sequencing have been obtained [8]. This strategy could potentially accelerate future genomic studies of *Wolbachia* and other endosymbiont bacteria that are difficult to purify in quantity.

Wolbachia are strictly intracellular, maternally transmitted, α-proteobacteria that infect many invertebrates including mites, filarial nematodes, crustaceans, spiders and at least 25% of all insect species [9, 10]. *Wolbachia* induce a number of phenotypes in their hosts that enhance their transmission and contribute to their success. In filarial nematodes, *Wolbachia* acts as an obligate mutualist, as removal of *Wolbachia* by antibiotics induces sterility, developmental defects and death of adult nematodes [11]. In arthropods, *Wolbachia* behave as reproductive parasites, and modify their host's reproduction in a variety of ways including male killing, feminization of genetic males, parthenogenesis induction within infected haplodiploid species or more commonly via cytoplasmic incompatibility (CI) [10]. All parasitic traits either increase the number, or provide a reproductive advantage to infected females, allowing *Wolbachia* to invade host populations even if a fitness cost is imposed on the host [12].

CI is the most widespread reproductive modification induced by *Wolbachia*. It is expressed when an infected male mates with a female that lacks the same strain of *Wolbachia* found in the male, resulting in failure to produce progeny. The reciprocal cross is fertile, as are crosses between males and females infected with the same strain of *Wolbachia*. Since infected females can mate successfully with either infected or uninfected males, while uninfected females are incompatible with infected males, they have a significant reproductive advantage [10]. Therefore, as a consequence of maternal transmission of *Wolbachia*, CI acts to replace uninfected with infected hosts in a given population. Although the molecular basis of how CI is induced is currently unknown, several lines of evidence suggest that *Wolbachia* infection disrupts the proper functioning of sperm [13]. Cytological studies demonstrate that nuclear envelope breakdown and mitosis are delayed during the first mitotic division associated with the expression of CI and consequent embryonic death in *Nasonia* [14].

Genome Features of *Wolbachia*

Comparison of the genome size of endosymbionts with related free-living bacteria reveals that bacterial endosymbionts generally have reduced genomes, typically around 1 Mb in size. The genome size reduction in obligate endosymbionts is thought to reflect a reduced requirement for many metabolic pathways within these bacteria, where the availability of metabolites from host cells has allowed the passive loss of genes associated with redundant pathways [15–17]. Recent experimental evidence suggests that genome streamlining can occur in a surprisingly short evolutionary time period [18]. Together with reduced genome size most endosymbiont genomes are very AT rich. This bias may reflect GC→AT mutational pressure due to the higher energy cost and limited availability of G and C in the host cell [19].

The two complete *Wolbachia* genomes, *w*Mel and *w*Bm are represented by a single circular chromosome consisting of 1.1–1.27 Mb with GC contents of about 35% [1, 2]. One of the most unusual features of insect *Wolbachia* genomes is the high content of repetitive DNA and their capacity for rearrangement. More than 14% of the *Wolbachia w*Mel chrosmosome consists of repetitive DNA, predominatly transposable element and prophage sequences [2]. Compared to *w*Mel, the mutualistic *Wolbachia w*Bm lacks many of these repetitive elements [1]. These sequences provide sites for genome rearrangements that have the potential to influence gene expression within the genome. This plasticity may be advantageous for *Wolbachia*, assisting its adaption to new hosts and facilitating invasion of new species. It also results in little gene colinearity between *Wolbachia* genomes examined to date.

Genome Rearrangements

Comparison of genome sequences from *w*Mel, *w*Ana and *w*Bm reveals a highly scrambled gene order. This suggests that *Wolbachia* chromosomes have been rearranged frequently through recombination [1, 2, 4]. The presence of functional DNA repair and recombination enzymes in the *Wolbachia* genome is consistent with the highly rearranged chromosomes [2]. Similar systems are known to promote genetic diversity of free-living bacterial genomes, by assisting in the insertion of foreign DNA or producing genomic rearrangements [20]. In contrast, most other obligate intracellular bacteria show conserved gene order structures, typified by different *Buchnera* endosymbionts of aphids, whose genomes have identical gene order structures despite at least 50 million years of divergence time [15, 21]. It is likely that the lack of DNA repair and recombination machinery together with intense purifying selection acts to reduce genomic instability or rearrangements in these obligate mutualists [22]. Recently, as a result of sequencing the *Rickettsia felis* genome the first putative conjugative plasmid in an obligate intracellular bacteria was identified. In addition, the *R. felis* genome seems to have some similarity to insect *Wolbachia* in that it contains high levels of repetitive DNA, as well as a disrupted gene order when compared to its relative *Rickettsia conorii* [7].

Mobile DNA

Mobile DNA in the form of transposons and bacteriophage are extremely common in prokaryote genomes. It has been estimated that up to 20% of some bacterial genomes are composed of bacteriophage genes [23]. The prokaryotes can be categorized into three groups based on their ecology. Free-living bacteria typically have large genomes (4–10 Mb) that contain a significant proportion of mobile DNA (~50%). Facultative intracellular bacteria, which invade the intracellular environment regularly while still maintaining the capacity for free-living growth, have intermediate genome sizes (2–7 Mb) with a typically moderate mobile DNA proportion (~12%). Obligate intracellular bacteria usually have small genomes (0.5–2 Mb) with quite small amounts of mobile DNA (~2%) [24]. Within these categories significant variation occurs. For example, some species of obligate intracellular bacteria show high levels of mobile DNA (e.g. *Wolbachia w*Mel has ~10%). Interestingly, mutualistic obligate intracellular species, such as *Blochmannia*, *Buchnera* and *Wigglesworthia* lack mobile DNA. In contrast, parasitic or pathogenic intracellular bacteria, such as *Chlamydia*, *Rickettsia* and *Wolbachia* contain much higher levels [24]. While the increased levels of mobile DNA may be an effect correlated with higher levels of horizontal transmission between hosts, it is also possible that mobile DNA may actually facilitate horizontal transmission through the acquisition of virulence genes and modulation of chromosomal gene expression. Such events would provide increased variation that may be selected during episodes of host switching. Insect *Wolbachia,* while predominantly vertically inherited, are known to have the capability to move between hosts from both phylogenetic analysis and experimental studies [25–27].

Transposable Elements

The most common form of mobile DNA in the *Wolbachia* genome are the transposable elements, comprising approximately 8% of the total coding sequences in *w*Mel [2]. Transposable elements can move from one locus to another, leading to replication within a host genome. In the *w*Mel genome, 13 copies of the IS5 transposable element exist and have identical sequences, suggesting that this element may be active [2]. Two major classes of transposable elements are represented in the *w*Mel genome, DNA transposable elements and retrotransposable elements. DNA transposable elements are identified by the presence of transposase-encoding genes. The *w*Mel genome contains a variety of these elements, such as the IS3, IS4, IS5, IS110, IS630 and ISBt12 families [2]. These families of DNA transposable elements can move by a conservative replicative process. They have little or no target specificity, and as such can have large mutagenic effects associated with insertion events [28] contributing to genetic diversity within symbiont lineages, and have disrupted at least nine genes within the *w*Mel genome [2].

The second type of transposable elements is the retrotransposon. These elements are identified by the presence of the reverse transcriptase-encoding gene and move through an RNA intermediate [28]. The *w*Mel genome contains four types of likely retrotransposable elements [2]. Three of them contain maturase domains, required for intron-specific splicing. A maturase domain in the reverse transcriptase gene is a strong feature of a mobile group II intron [24]. Mobile group II introns are self-splicing mobile retroelements, can insert themselves into both specific (homing sites) and ectopic sites (nonhomologous sites) by reverse splicing. They are found in bacteria and in eukaryotic organelle genomes, such as mitochondria and chloroplasts [29]. They are also of interest because they are the putative ancestors of the spliceosome-dependent eukaryotic nuclear introns [30]. The recent comparative genomics data suggest substantial horizontal transfer of these introns among different bacterial species [31] potentially through conjugation mechanisms [32]. The phylogenetic relationships among mobile group II introns indicate that they evolved in bacteria and then were likely transferred to eukaryotes, via organelles that originated from bacterial endosymbionts [31]. The role of mobile group II introns in *Wolbachia* genomes is unclear; however, they have the potential to be applied as part of transformation technology to determine gene function via gene knockout experiments.

Prophage

Bacteriophages, viruses of prokaryotes, are one of the most effective vehicles for horizontal transfer of foreign DNA into recipient bacteria and are agents known to increase genome diversification [23]. Prophages are integrated genomes of bacteriophages that they are passively inherited. Prophage sequences are commonly observed in free-living and facultative intracellular bacteria. In contrast, obligate intracellular bacteria typically have no or few prophage sequences [24]. The absence of prophage genes in *Wigglesworthia* and *Buchnera*, mutualistic obligate intracellular γ-proteobacteria, as well as *w*Bm, the mutualistic *Wolbachia* that infects the filarial nematode, reflects the evolutionary pressure to remove nonessential genes, leading to genome size reduction. Approximately half the completed prokaryotic genome sequences do not contain any prophage-related genes [33], suggesting that they are likely to be eliminated if not adaptive. The genomes of parasitic, obligate, intracellular bacteria often contain a limited number of prophage genes. *Chlamydia* and some *Rickettsia* strains for example, have only one prophage gene. However, the *w*Mel genome contains 28 genes of phage origin, their function being largely unknown. Virulence genes of pathogenic bacteria are commonly located in chromosomal prophage regions [34]. Where these genes are retained in degraded prophage regions it is thought that the functions of these genes have been co-opted by the bacterial host [33]. The presence of prophages in parasitic obligate intracellular bacteria suggests a role for these genes in the parasitic ecology of these *Wolbachia* strains.

The observation of phage particles from several strains of *Wolbachia* [35–37], and the simultaneous infection of different *Wolbachia* strains in the same host cell [38] suggests that phage are capable of facilitating the horizontal transfer of genes among different *Wolbachia* lineages. This is supported by recent phylogenetic analysis of the phage gene *orf7*, a putative capsid protein of phage WO, which revealed that distantly related *Wolbachia* strains contain closely related prophage gene sequences [39, 40].

WO bacteriophage particles have been successfully isolated from *w*Cau-infected flour moth cells and the WOcauB1 phage genome sequence determined. This phage genome contains 24 open reading frames that encode for capsid proteins, DNA packaging proteins, phage particle assembly proteins and several proteins of unknown function. Of particular interest are genes encoding the hypothetical secreted proteins Gp15 and Gp16, similar to VrlC, a virulence-related protein [35]. Interestingly, these proteins are also found in *w*Mel prophage regions.

*w*Mel contains three prophage regions, two of which are closely related to phage WO, named WO-A and WO-B [2]. WO-B was originally thought to be inactive due to the comparison between WO-B and *w*Kue which showed a major rearrangement and translocation. However, phage-like particles containing the *orf7* gene in filtered extracts from *w*Mel-infected *D. melanogaster* [40] suggest otherwise. This contradiction might relate to the recent discovery that the *w*Mel strain of *Wolbachia* is actually composed of a number of distinct genetic strains [41], some of which may contain active phage [Riegler, pers. commun.]. A number of the phage-associated genes in the *w*Mel genome encode for ankyrin domain (ANK) proteins [2]. They are notable because they are known to mediate protein-protein interactions in eukaryotic cells, and have the potential to interact with host proteins. Furthermore, the related intracellular pathogen, *Anaplasma phagocytophilum* secretes an ANK protein (AnkA) into host cells that binds to host chromatin and is thought to regulate host gene expression [42]. Recent studies have identified a correlation between ANK gene and reproductive manipulation phenotype by comparing ANK gene sequence variations in *w*Mel and *w*Au, a closely related non-CI-inducing strain from *Drosophila simulans* [43]. The phage WO-associated ANK gene WD0636 harbors a point mutation that results in the loss of one ANK motif, and WD0633 has a small insertion and deletion of amino acids in *w*Au. These mutations could potentially affect its function and correlate with CI phenotype. Independently, the relationship between ANK gene variation and CI expression has been shown in *w*Pip, a *Wolbachia* strain that infects the mosquito *C. pipiens* [37]. Substantial divergence in two ANK genes (*pk1* and *pk2*), located in the prophage WO region are found in two *Wolbachia* strains, which are also bidirectionally incompatible. Moreover, expression of *pk2* was only seen in female mosquitoes [37]. Similar sex-specific expression of the *orf7* gene has been shown in different *Culex* strains [36], suggesting that the expression of phage WO genes is differentially regulated in *Culex* males and females. Because CI mechanisms are thought to be related to sperm modification in males and fertilization rescue in females, the presence of sex-specific expression of these genes is an interesting observation.

In the *C. pipiens* mosquito group, very complex CI patterns have been reported; however, no *Wolbachia* strain variation can be found using sequences of *ftsZ, 16S rRNA* and *wsp* genes [44–46]. A straightforward hypothesis to explain the complexity of CI pattern is the presence of extrachromosomal genetic factors, such as transposable elements and phages, which contribute in some way to determining CI pattern in *Culex*. In support of this hypothesis, significant polymorphism has been found in phage-related *pk* genes [37] and *orf7* [47], as well as an IS5 family transposable element *Tr1* [44]. Although a causal relationship between polymorphic markers and CI pattern has not been identified, phage-related genes and transposable elements are potential candidates to provide an explanation for the complex CI patterns seen in this species and ultimately an understanding of CI mechanisms.

***Wolbachia* Genomics and Population Biology**

One practical consequence of the genomic plasticity that insect *Wolbachia* display is the resulting abundance of highly polymorphic markers that can readily discriminate between closely related strains. Previously, *Wolbachia* strains have been distinguished by the sequence variation of several different genetic loci including *16S rRNA, ftsZ, dnaA* and *wsp* gene sequences [38, 48–50]. However, in the *Wolbachia* that infect *D. melanogaster*, these genes all have essentially identical sequences between different *D. melanogaster* populations, indicating that this insect host species is infected by a single strain of *Wolbachia*, known as *w*Mel. However, considerable variation in CI levels have been observed in different populations of *D. melanogaster* infected with *w*Mel [51]. This phenotypic variability has been thought to be influenced by different host genetic backgrounds [52, 53].

The completed *w*Mel genome sequence and subsequent analysis revealed a number of novel polymorphic genetic markers, including the presence of IS5 transposable element insertion variability and size variation in variable number tandem repeat regions. Using these new markers it was possible to distinguish five *w*Mel variants within different *D. melanogaster* populations [41]. The discovery of five *w*Mel variants in what was originally thought to be an essentially clonal symbiont lineage suggests that the phenotypic variability previously described is probably associated with the different properties of these variants. Interestingly, analysis of long-term lab stocks and museum specimens show that one of these variants has replaced all other variants in field populations within the last century. This sweep suggests the presence of differential fitness effects between these different strains, which has contributed to the appearance of one dominant variant. Although CI has been proposed as the driving force for population replacement, the persistent infection of *D. melanogaster* with *Wolbachia* is hard to explain due to the weak levels of CI expressed in field populations [54]. It has been speculated that *w*Mel might provide positive fitness effects to *D. melanogaster*, which may explain the current *w*Mel dominance; however, the nature of the benefit conferred by *Wolbachia* to the insect host has yet to be found [55]. Recent analysis of the *w*Bm genome indicates that metabolic provisioning may form part of the observed mutualism between *Wolbachia* and its nematode host [1]. Intriguingly, these same pathways are present within insect *Wolbachia* lineages and may form the basis of a benefit to insects during periods of nutritional stress [Brownlie, pers. commun.].

Nonneutral Evolution in *Wolbachia* Genomes

Three-way genome comparisons have been employed previously to identify genes experiencing differential patterns of nonneutral evolution in two closely related lineages [56]. Recently, Brownlie et al. [57] have applied this comparative approach to the genomes of *w*Mel and *w*Bm relative to that of *Anaplasma marginale*. Signatures of selection as measured by per gene estimates of d_N/d_S ratios were calculated for 591 orthologous gene sets across the three genomes. Genes with ratios significantly greater than one reveal a past history of diversifying selection. Using genome annotation information regarding functional class, the involvement of selection in different biological processes in the symbiont lineages could be evaluated. The overall goals of the study were to identify candidate genes underlying (1) *Wolbachia* specific adaptations relative to *Anaplasma* and (2) *Wolbachia* host specific adaptations to *B. malayi* versus *D. melanogaster*. Although many individual functional categories of genes demonstrated strong evidence of selection in the *Wolbachia* genomes, there were also clear overarching and re-emerging trends in terms of symbiont biology. A few of these trends are summarized below:

The Synthesis of Cofactors

The basis of *Wolbachia's* dependence on its host and the nature of any benefits *Wolbachia* might provide to the filarial nematode and to insects are two fundamental unanswered questions in *Wolbachia* biology. The role of cofactors in the dependence of insect and endosymbiont partners on one another has been hypothesized for other associations [58]. The complete genome sequence analyses of *w*Mel [2] and *w*Bm [1] revealed incomplete cofactor biosynthetic pathways in *Wolbachia* and the presence of import/export pathways not present in close phylogenetic relatives of *Wolbachia*. The genome analysis and that of Brownlie et al. [57] both support the involvement of iron as an interaction point between host and microbe. Both genomes contain pathways for heme biosynthesis [1, 2], and it appears that the nematode *B. malayi* may be incapable of synthesizing heme. Multiple gene members of the heme biosynthetic pathway have experienced selection in the two *Wolbachia* lineages, as have 2 heme exporters in *w*Mel and 2 iron uptake transporters in *w*Bm. Although insect hosts are not dependent on *Wolbachia* for heme biosynthesis, the microbe may supplement host stores or play an additional role in iron homeostasis.

Secretion

For an intracellular microbe, secretion represents the main route of communication with the host and extracellular environment. The selection analysis of Brownlie et al. [57] indicates that 11 genes involved with secretion in *w*Mel and 8 in *w*Bm have experienced accelerated evolution and selection. At least 3 of the genes in the type IV pathway, which facilitates host-endosymbiont communication for numerous systems [59], are selected in *w*Bm.

Replication

The coordination of symbiont and host cell replication rates is a clear requirement that occurs through unknown mechanisms. In the case of *Wolbachia*, the symbiont has to be able to deal with the challenges of insect diapause [60] and selection against symbiont cancers [10, 61], as demonstrated by the *D. melanogaster* *w*MelPop strain. Multiple genes in both symbiont lineages that are involved with cell division, particularly with regulation of growth rates, stress and stationary phase appear to be affected by selection. This theme of selection on growth processes is mirrored by genes involved with DNA replication enzymes, the process of translation, and peptidoglycan synthesis.

Protein Stability

The pioneering work identifying the accumulation of slightly deleterious mutants in *Buchnera* [22] has predicted the importance of chaperones like GroEL in maintaining the integrity of proteins in symbionts [62]. The 3-way genome selection analysis has detected diversifying selection in GroEL in both *Wolbachia* genomes and selection on at least 6 other genes in *w*Mel that contribute to proper protein folding and stability. The greater emphasis of this feature in the *w*Mel genome is intriguing and may point to variation in bottleneck size in the different hosts. Alternatively, *w*Bm and *w*Mel may have evolved different strategies for coping with the mutational pressures associated with small population size.

References

1 Foster J, Ganatra M, Kamal I, Ware J, Makarova K, Ivanova N, Bhattacharyya A, Kapatral V, Kumar S, Posfai J, et al: The *Wolbachia* genome of *Brugia malayi*: endosymbiont evolution within a human pathogenic nematode. PLoS Biol 2005;3:e121.

2 Wu M, Sun LV, Vamathevan J, Riegler M, Deboy R, Brownlie JC, McGraw EA, Martin W, Esser C, Ahmadinejad N, et al: Phylogenomics of the reproductive parasite *Wolbachia pipientis* wMel: a streamlined genome overrun by mobile genetic elements. PLoS Biol 2004;2:E69.

3 Iturbe-Ormaetxe I, Riegler M, O'Neill SL: New names for old strains? *Wolbachia* wSim is actually wRi. Genome Biol 2005;6:401; author reply 401.

4 Salzberg SL, Hotopp JC, Delcher AL, Pop M, Smith DR, Eisen MB, Nelson WC: Serendipitous discovery of *Wolbachia* genomes in multiple *Drosophila* species. Genome Biol 2005;6:R23.

5 Andersson SG, Zomorodipour A, Andersson JO, Sicheritz-Ponten T, Alsmark UC, Podowski RM, Naslund AK, Eriksson AS, Winkler HH, Kurland CG: The genome sequence of *Rickettsia prowazekii* and the origin of mitochondria. Nature 1998;396:133–140.

6 McLeod MP, Qin X, Karpathy SE, Gioia J, Highlander SK, Fox GE, McNeill TZ, Jiang H, Muzny D, Jacob LS, et al: Complete genome sequence of *Rickettsia typhi* and comparison with sequences of other rickettsiae. J Bacteriol 2004;186:5842–5855.

7 Ogata H, Renesto P, Audic S, Robert C, Blanc G, Fournier PE, Parinello H, Claverie JM, Raoult D: The genome sequence of *Rickettsia felis* identifies the first putative conjugative plasmid in an obligate intracellular parasite. PLoS Biol 2005;3:e248.

8 Mavingui P, Van VT, Labeyrie E, Rances E, Vavre F, Simonet P: Efficient procedure for purification of obligate intracellular *Wolbachia pipientis* and representative amplification of its genome by multiple-displacement amplification. Appl Environ Microbiol 2005;71:6910–6917.

9 Jeyaprakash A, Hoy MA: Long PCR improves *Wolbachia* DNA amplification: wsp sequences found in 76% of sixty-three arthropod species. Insect Mol Biol 2000;9:393–405.

10 Werren JH, O'Neill SL: The evolution of heritable symbionts; in O'Neill SL, Hoffmann AA, Werren JH (eds): Influential Passengers: Inherited Microorganisms and Arthropod Reproduction. Oxford, Oxford University Press, 1997.

11 Taylor M: One in the eye for river blindness. Trends Microbiol 2001;9:310.

12 Hoffmann AA, Turelli M: Cytoplasmic incompatibility in insects; in O'Neill SL, Hoffmann AA, Werren JH (eds): Influential Passengers: Inherited Microorganisms and Arthropod Reproduction. Oxford, Oxford University Press, 1997.
13 Werren JH: Biology of *Wolbachia*. Annu Rev Entomol 1997;42:587–609.
14 Tram U, Sullivan W: Role of delayed nuclear envelope breakdown and mitosis in *Wolbachia*-induced cytoplasmic incompatibility. Science 2002;296:1124–1126.
15 Klasson L, Andersson SG: Evolution of minimal-gene-sets in host-dependent bacteria. Trends Microbiol 2004;12:37–43.
16 Sallstrom B, Andersson SG: Genome reduction in the alpha-proteobacteria. Curr Opin Microbiol 2005;8:579–585.
17 Wernegreen JJ: For better or worse: genomic consequences of intracellular mutualism and parasitism. Curr Opin Genet Dev 2005;15:572–583.
18 Nilsson AI, Koskiniemi S, Eriksson S, Kugelberg E, Hinton JC, Andersson DI: Bacterial genome size reduction by experimental evolution. Proc Natl Acad Sci USA 2005;102:12112–12116.
19 Rocha EP, Danchin A: Base composition bias might result from competition for metabolic resources. Trends Genet 2002;18:291–294.
20 Lorenz MG, Wackernagel W: Bacterial gene transfer by natural genetic transformation in the environment. Microbiol Rev 1994;58:563–602.
21 Tamas I, Klasson L, Canback B, Naslund AK, Eriksson AS, Wernegreen JJ, Sandstrom JP, Moran NA, Andersson SG: 50 million years of genomic stasis in endosymbiotic bacteria. Science 2002;296:2376–2379.
22 Moran NA: Accelerated evolution and Muller's rachet in endosymbiotic bacteria. Proc Natl Acad Sci USA 1996;93:2873–2878.
23 Casjens S: Prophages and bacterial genomics: what have we learned so far? Mol Microbiol 2003;49:277–300.
24 Bordenstein SR, Reznikoff WS: Mobile DNA in obligate intracellular bacteria. Nat Rev Microbiol 2005;3:688–699.
25 Boyle L, O'Neill SL, Robertson HM, Karr TL: Interspecific and intraspecific horizontal transfer of *Wolbachia* in *Drosophila*. Science 1993;260:1796–1799.
26 Sinkins SP, Braig HR, O'Neill SL: *Wolbachia* superinfections and the expression of cytoplasmic incompatibility. Proc Biol Sci 1995;261:325–330.
27 Xi Z, Khoo CC, Dobson SL: *Wolbachia* establishment and invasion in an Aedes aegypti laboratory population. Science 2005;310:326–328.
28 Labrador M, Corces VG: Transposable element-host interactions: regulation of insertion and excision. Annu Rev Genet 1997;31:381–404.
29 Lambowitz AM, Zimmerly S: Mobile group II introns. Annu Rev Genet 2004;38:1–35.
30 Sharp PA: Five easy pieces. Science 1991;254:663.
31 Zimmerly S, Hausner G, Wu X: Phylogenetic relationships among group II intron ORFs. Nucleic Acids Res 2001;29:1238–1250.
32 Belhocine K, Plante I, Cousineau B: Conjugation mediates transfer of the Ll.LtrB group II intron between different bacterial species. Mol Microbiol 2004;51:1459–1469.
33 Canchaya C, Fournous G, Brussow H: The impact of prophages on bacterial chromosomes. Mol Microbiol 2004;53:9–18.
34 Waldor MK, Friedman DI: Phage regulatory circuits and virulence gene expression. Curr Opin Microbiol 2005;8:459–465.
35 Fujii Y, Kubo T, Ishikawa H, Sasaki T: Isolation and characterization of the bacteriophage WO from *Wolbachia*, an arthropod endosymbiont. Biochem Biophys Res Commun 2004;317:1183–1188.
36 Sanogo YO, Dobson SL: WO bacteriophage transcription in *Wolbachia*-infected *Culex pipiens*. Insect Biochem Mol Biol 2006;36:80–85.
37 Sinkins SP, Walker T, Lynd AR, Steven AR, Makepeace BL, Godfray HC, Parkhill J: *Wolbachia* variability and host effects on crossing type in *Culex* mosquitoes. Nature 2005;436:257–260.
38 Werren JH, Zhang W, Guo LR: Evolution and phylogeny of *Wolbachia*: reproductive parasites of arthropods. Proc Biol Sci 1995;261:55–63.
39 Bordenstein SR, Wernegreen JJ: Bacteriophage flux in endosymbionts (*Wolbachia*): infection frequency, lateral transfer, and recombination rates. Mol Biol Evol 2004;21:1981–1991.

40 Gavotte L, Vavre F, Henri H, Ravallec M, Stouthamer R, Bouletreau M: Diversity, distribution and specificity of WO phage infection in *Wolbachia* of four insect species. Insect Mol Biol 2004;13: 147–153.

41 Riegler M, Sidhu M, Miller WJ, O'Neill SL: Evidence for a global *Wolbachia* replacement in *Drosophila melanogaster*. Curr Biol 2005;15:1428–1433.

42 Caturegli P, Asanovich KM, Walls JJ, Bakken JS, Madigan JE, Popov VL, Dumler JS: AnkA: an *Ehrlichia phagocytophila* group gene encoding a cytoplasmic protein antigen with ankyrin repeats. Infect Immun 2000;68:5277–5283.

43 Iturbe-Ormaetxe I, Burke GR, Riegler M, O'Neill SL: Distribution, expression, and motif variability of ankyrin domain genes in *Wolbachia pipientis*. J Bacteriol 2005;187:5136–5145.

44 Duron O, Lagnel J, Raymond M, Bourtzis K, Fort P, Weill M: Transposable element polymorphism of *Wolbachia* in the mosquito *Culex pipiens*: evidence of genetic diversity, superinfection and recombination. Mol Ecol 2005;14:1561–1573.

45 Guillemaud T, Pasteur N, Rousset F: Contrasting levels of variability between cytoplasmic genomes and incompatibility types in the mosquito *Culex pipiens*. Proc Biol Sci 1997;264: 245–251.

46 Rousset F, Bouchon D, Pintureau B, Juchault P, Solignac M: *Wolbachia* endosymbionts responsible for various alterations of sexuality in arthropods. Proc Biol Sci 1992;250:91–98.

47 Sanogo YO, Dobson SL: Molecular discrimination of *Wolbachia* in the *Culex pipiens* complex: evidence for variable bacteriophage hyperparasitism. Insect Mol Biol 2004;13:365–369.

48 Bourtzis K, Nirgianaki A, Onyango P, Savakis C: A prokaryotic dnaA sequence in *Drosophila melanogaster*: *Wolbachia* infection and cytoplasmic incompatibility among laboratory strains. Insect Mol Biol 1994;3:131–142.

49 Holden PR, Jones P, Brookfield JF: Evidence for a *Wolbachia* symbiont in *Drosophila melanogaster*. Genet Res 1993;62:23–29.

50 Zhou W, Rousset F, O'Neil S: Phylogeny and PCR-based classification of *Wolbachia* strains using wsp gene sequences. Proc Biol Sci 1998;265:509–515.

51 Mercot H, Charlat S: *Wolbachia* infections in *Drosophila melanogaster* and *D. simulans*: polymorphism and levels of cytoplasmic incompatibility. Genetica 2004;120:51–59.

52 Reynolds KT, Hoffmann AA: Male age, host effects and the weak expression or non-expression of cytoplasmic incompatibility in *Drosophila* strains infected by maternally transmitted *Wolbachia*. Genet Res 2002;80:79–87.

53 Reynolds KT, Thomson LJ, Hoffmann AA: The effects of host age, host nuclear background and temperature on phenotypic effects of the virulent *Wolbachia* strain popcorn in *Drosophila melanogaster*. Genetics 2003;164:1027–1034.

54 Hoffmann AA, Hercus M, Dagher H: Population dynamics of the *Wolbachia* infection causing cytoplasmic incompatibility in *Drosophila melanogaster*. Genetics 1998;148:221–231.

55 Harcombe W, Hoffmann AA: *Wolbachia* effects in *Drosophila melanogaster*: in search of fitness benefits. J Invertebr Pathol 2004;87:45–50.

56 Clark AG, Glanowski S, Nielsen R, Thomas PD, Kejariwal A, Todd MA, Tanenbaum DM, Civello D, Lu F, Murphy B, et al: Inferring nonneutral evolution from human-chimp-mouse orthologous gene trios. Science 2003;302:1960–1963.

57 Brownlie JB, Adamski M, Slatko B, McGraw EA: Nonneutral evolution in *Wolbachia* genomes. PLoS Biol, submitted.

58 Rio RV, Lefevre C, Heddi A, Aksoy S: Comparative genomics of insect-symbiotic bacteria: influence of host environment on microbial genome composition. Appl Environ Microbiol 2003;69: 6825–6832.

59 Christie PJ, Atmakuri K, Krishnamoorthy V, Jakubowski S, Cascales E: Biogenesis, architecture, and function of bacterial type IV secretion systems. Annu Rev Microbiol 2005;59:451–485.

60 Ruang-areerate T, Kittayapong P, McGraw EA, Baimai V, O'Neill SL: *Wolbachia* replication and host cell division in *Aedes albopictus*. Curr Microbiol 2004;49:10–12.

61 McGraw EA, Merritt DJ, Droller JN, O'Neill SL: *Wolbachia* density and virulence attenuation after transfer into a novel host. Proc Natl Acad Sci USA 2002;99:2918–2923.

62 Fares MA, Moya A, Barrio E: Adaptive evolution in GroEL from distantly related endosymbiotic bacteria of insects. J Evol Biol 2005;18:651–660.

S.L. O'Neill, PhD
School of Integrative Biology, The University of Queensland, St. Lucia
QLD 4072, Brisbane (Australia)
Tel. +61 7 33652471, Fax +61 7 33469213, E-Mail scott.oneill@uq.edu.au

Hoerauf A, Rao RU (eds): Wolbachia.
Issues Infect Dis. Basel, Karger, 2007, vol 5, pp 90–123

Wolbachia Symbiosis in Arthropods

Michael E. Clark

Department of Biology, University of Rochester, Rochester, N.Y., USA

Abstract

Wolbachia pipientis is an intracellular bacteria found within the cytoplasm of a high proportion of arthropods. Widespread in insects, *Wolbachia* is also commonly found in other arthropod groups, including mites, spiders and terrestrial isopods. *Wolbachia* are normally maternally inherited and have evolved a number of strategies to ensure transmission. These include: (1) feminization, the conversion of genetic males into females, (2) parthenogenesis, the production of diploid offspring without sexual reproduction, (3) male killing, the killing of infected males to the benefit of infected female siblings, and (4) cytoplasmic incompatibility, the inability of infected males to successfully fertilize eggs from either uninfected females or females infected with different *Wolbachia* types. In addition to this reproductive parasitism, *Wolbachia* can influence other aspects of host fitness, including host longevity, fecundity, fertility and host-parasitoid interactions. *Wolbachia* has been studied extensively in the context of host spermatogenesis, oogenesis and embryogenesis. These cytological studies of host-*Wolbachia* interactions are providing insights into the ways in which *Wolbachia* are able to manipulate hosts.

The genus *Wolbachia* is a widespread group of intracellular bacteria found in arthropods and filarial nematodes [1, 2]. The genus occupies a position intermediate between the helminth-borne *Neorickettsia* group and the tick-transmitted groups *Ehrlichia* and *Anaplasma* [3]. Whereas members of *Neorickettsia, Ehrlichia* and *Anaplasma* are known to infect vertebrates, either incidentally or as a normal part of their life cycle, *Wolbachia* have so far only been found as intracellular bacteria associated with arthropods and filarial nematodes [3]. In contrast, *Wolbachia* are routinely transmitted within a host species by vertical transmission through the egg cytoplasm of their hosts, although on larger time scales they show horizontal transmission between host species, accounting for their widespread distribution [3].

For decades, *Wolbachia* was known only from mosquitoes [4, 5]. With the development of PCR-based screening methods, it has become clear that *Wolbachia* in arthropods is widespread in nature. Several surveys have suggested that around 20% of arthropods are infected with *Wolbachia* [1, 6, 7]. Other surveys have found as many as many as 76% of arthropods in the community surveyed were infected [8]. This makes *Wolbachia* among the most common intracellular bacteria known, with estimates of several million infected species [6]. Although found in all major insect orders, *Wolbachia* are not evenly distributed. In some groups, such as lice, *Wolbachia* seems ubiquitous [9], while in others, such as Anopheles mosquitoes, *Wolbachia* is completely absent [10–12]. Widespread surveys of arthropods are currently underway to determine global *Wolbachia* infection frequencies.

Wolbachia are remarkable in their ability to manipulate the cell and reproductive biology of their hosts. The most commonly described phenotypes are (1) feminization of genetic males, (2) parthenogenesis, (3) male killing, and (4) cytoplasmic incompatibility (CI), a form of conditional infertility. Each of these phenotypes increases the frequency of infected females in the host population, and therefore are bacterial adaptations to increase transmission of the microorganisms. Such parasite effects on hosts are commonly referred to as 'reproductive parasitism' [13]. *Wolbachia* are also known to effect hosts in a number of other ways, for example in different host species removal of *Wolbachia* can inhibit ovarian development, influence host longevity, or alter host-parasitoid interactions. Each of these phenotypes will be discussed in detail below.

Phenotypic Effects of *Wolbachia* on Their Hosts

Koch's postulates are impossible to strictly apply to *Wolbachia*, as well as other endosymbiotic bacteria, due to the current impossibility of culturing *Wolbachia* in a cell-free environment. Since most bacteria (pathogenic, symbiotic and others) are unculturable, alternatives to Koch's postulates must be used to assess the effects of such bacteria on hosts [14]. Many different levels of evidence have been used to attribute the phenotypic effects to *Wolbachia* infections. Obviously, the weakest line of evidence is simply the presence of a phenotype and *Wolbachia* (for example, see the correlative case of sex ratio distortion and *Wolbachia* infection in the human head louse [15]). A somewhat stronger case can be made when a phenotype is observed only within *Wolbachia*-infected individuals and is absent in all *Wolbachia*-free individuals tested within a species. This approach alone is not definitive because other microorganisms may be coinfected with *Wolbachia* and responsible for the phenotype in question. Ideally, a thorough search for other symbionts should accompany any such analysis. Other microorganisms with effects similar to *Wolbachia* will be discussed later. In addition to microorganisms, host genetic factors may cause *Wolbachia*-like phenotypes (sex ratio distortion, parthenogenesis, male sterility).

Thorough investigations of the phenotypic effects of *Wolbachia* require a level of experimental manipulation. A more rigorous investigation of the effects of *Wolbachia* on a host must include a phenotypic analysis in both the presence and absence of *Wolbachia*, within the same host nuclear background. This approach has been the standard for most investigations into *Wolbachia* effects. This requires the removal of *Wolbachia*, which in most cases can easily be accomplished by antibiotic or heat treatments. In many cases, it has been possible to create new *Wolbachia* infections in previously uninfected hosts, or reestablishment *Wolbachia* infections in cured lines using microinjection techniques [16–18]. Obviously, such investigations have been limited to host organisms which can be reared under laboratory conditions. Consequently, much of what is currently known regarding the phenotypic effects of *Wolbachia* come from previously established model systems (*Culex, Drosophila, Nasonia, Tribolium*). For further discussion on the methods necessary for attributing specific phenotypes to a specific endosymbiont, see Weeks et al. [19].

Due to the aforementioned requirements, although *Wolbachia* has now been detected in hundreds of different arthropod species, the effects of *Wolbachia* infection on hosts have been investigated in only a relatively small number. Nevertheless, with these caveats in mind, considerable progress has been made in elucidating the effects of *Wolbachia*. The following is a summary of what is currently known.

Feminization

Feminization is perhaps the most obviously beneficial strategy for a maternally inherited bacterium such as *Wolbachia*. With males being dead-ends for the inheritance of most cytoplasmic factors (such as *Wolbachia* and mitochondria), conversion of infected genetic male offspring into females doubles the potential *Wolbachia* transmission to the following generation (fig. 1). To date however, *Wolbachia*-induced feminization is the most infrequently described of the four main *Wolbachia*-induced phenotypes, reported in only two Arthropod orders. *Wolbachia*-induced feminization has been documented most commonly in several species of terrestrial isopod within the order Oniscidae. The presence of a cytoplasmic microorganism was first described in association with feminization in *Armadillidium vulgare* in 1973 [20], and later identified as *Wolbachia*, causing genetic males to develop into females [21]. In addition to *A. vulgare, Wolbachia*-induced feminization in isopods has now been described in *Armadillidium pulchellum, Porcellionides pruinosus*, *Oniscus asellus* [22–25].

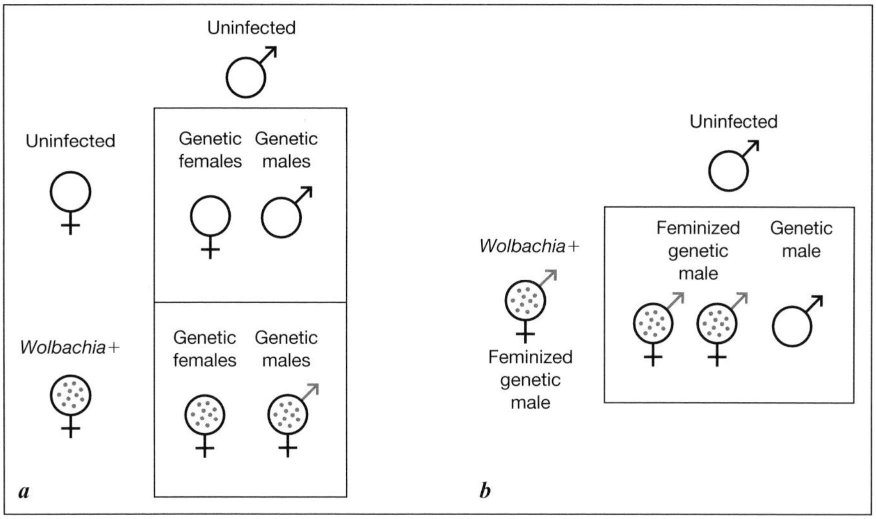

Fig. 1. Feminization. ***a*** Feminizing *Wolbachia* within mixed infected and uninfected populations. Three types of females are present within populations: uninfected genetic females, infected genetic females and infected feminized genetic males. Only one male type is seen, uninfected genetic males. ***b*** Feminizing *Wolbachia* in a population entirely consisting of genetic males. All females are infected feminized genetic males. Uninfected males are generated each generation by imperfect *Wolbachia* transmission from uninfected mothers.

In these isopod hosts, *Wolbachia* within genetic males inhibits the development of the androgenic gland and resulting androgenic hormone [26]. The resulting genetic males develop as females by default. These 'feminized' males, may however suffer a fitness disadvantage compared to genetic females, with males preferring to mate with, and transfer more sperm to, genetic females [27]. In addition, feminization is sometimes incomplete, with a mixture of male and female morphology found in *Wolbachia*-infected genetic males [25].

One obvious result of feminizing *Wolbachia* can be a dearth of males in an infected population. As a result of the biased sex ratio, males from populations with feminizing *Wolbachia* have a higher mating capacity than those from populations with only CI *Wolbachia* [28]. To date, all of the *Wolbachia* known to, or suspected to, induce feminization in isopods belong to the B group of *Wolbachia*, although not forming a monophyletic clade [29]. Within *A. vulgare* alone, two distinct *Wolbachia* strains have been reported to induce feminization, suggesting that this *Wolbachia* trait has evolved multiple times [24].

Among insects, *Wolbachia*-induced feminization is only currently known from a single *Wolbachia* strain within a single Lepidopterian species, *Eurema hecabe* [30]. Removal of *Wolbachia* from *E. hecabe* results in all male broods. Examination of nuclei from *Wolbachia*-infected females indicated that they lack the normal sex chromatin body corresponding to the W chromosome usually found in genetic females. All *Wolbachia*-infected females therefore are likely genetic males. With Lepidopterians lacking a sex-determining hormone similar to, or even analogous to, the androgenic hormone from isopods, it is likely that the *Wolbachia*-induced feminization in isopods is achieved through a different mechanism than that seen in *E. hecabe*.

The first report of suspected *Wolbachia*-induced feminization in insects was described in another Lepidopterian, the adzuki bean borer, *Ostrinia scapulalis*. All female broods were correlated with *Wolbachia* infection. Subsequent removal of *Wolbachia* via tetracycline treatment resulted in all male broods [31]. Further work however indicated that the all-female broods associated with *Wolbachia* infection were actually accomplished via male killing. While genetic males (ZZ) died as larva when infected with *Wolbachia*, uninfected females (WZ) created by tetracycline treatment died when uninfected. Genetic females from *Wolbachia*-infected lines were dependent on *Wolbachia* for survival. Males then cannot normally live with *Wolbachia*, while females cannot live without it [32]. Viable, infected genetic males were generated using two different methods, microinjection of *Wolbachia* into uninfected embryos, and mild tetracycline treatment of *Wolbachia*-infected mothers prior to oviposition. These infected genetic males, with sublethal doses of *Wolbachia*, developed as sexual mosaics, with phenotypically male and female tissues. Complete feminization of genetic males then, as normally seen with natural occurring levels of *Wolbachia* was lethal. Therefore, *Wolbachia*-induced male killing in *O. scapulalis* is accomplished via lethal feminization of genetic males, likely through the manipulation of some component of sex determination [32]. This work is worth noting not only to highlight the way in which one *Wolbachia*-induced phenotype (male killing) may easily be mistaken for another (feminization), but also to emphasize the similar mechanisms (feminization) which may lead to these different phenotypes (both feminization and male killing). In this host-*Wolbachia* association, there is not a clear distinction between feminization and male killing, either phenotypically of mechanistically.

As with many selfish genetic elements, the complete spread of a feminizing *Wolbachia* could be catastrophic for the host population. Fixation of a feminizing *Wolbachia* infection with complete penetrance would eliminate phenotypic males and lead to the extinction of both the host population as well as *Wolbachia*. A single generation of only females would obviously be ultimately disadvantageous for both host and symbiont, leading to the extinction of both.

Such events may occur in nature, but would for obvious reasons not be readily observed. Instead, those systems that do persist do so because of some mechanism of constraint on the ability of such a *Wolbachia* to spread. One mechanism is local scarcity of males resulting in reduced insemination of females in populations with higher infection levels. This process is termed 'group' or 'deme' selection, and requires some form of subdivision in host populations. At low levels within a population, females (both genetic females and feminized genetic males) can mate with uninfected males, which are the sons of uninfected females (fig. 1a). When a feminizing *Wolbachia* reaches high frequencies within a local population however, males can become a limiting factor. This selection pressure is likely the reason for a higher mating capacity of (uninfected) males from populations with feminizing *Wolbachia* infections [28]. Another resolution to this same dilemma is seen in the isopod *O. asellus*, where the transmission is far from perfect, below that seen with all other known naturally occurring *Wolbachia* infections. Transmission of *Wolbachia* in *O. asellus* is usually below 88% of offspring, assuring the production of (uninfected) males each generation [23]. In these populations, most individuals are actually genetically male, so loss of *Wolbachia* produces mostly phenotypic males (fig. 1b). Finally, selection for host genes which masculinize infected males are known to occur [28].

Feminizing *Wolbachia* infections are notably absent from species in which males are the heterogametic sex and expression of sex-linked genes are under dosage compensation. These infections are not likely to occur due to the likely lethality of feminizing such males – these infections would therefore be phenotypically male killing and only spread under certain conditions (discussed below). Even in the absence of dosage compensation, feminization of heterogametic genetic males would likely necessitate high levels of lethality (25%) in their offspring due to the effect of homozygosity of Y chromosomes. We would predict that *Wolbachia*-induced feminization will not be found in groups like dipterans where male heterogamety is the rule. Therefore, the limits to the distribution of feminizing *Wolbachia* infections include both sex determining mechanisms and heterogamety.

Parthenogenesis

With males being an evolutionary dead end for *Wolbachia* inheritance, another obvious strategy of host manipulation by a maternally inherited endosymbiont is to induce parthenogenesis, the production of female offspring without fertilization by sperm (fig. 2). As with *Wolbachia*-induced feminization, parthenogenesis induction doubles the potential transmission of *Wolbachia* to the next generation, because all the progeny are female. Currently, *Wolbachia*-induced parthenogenesis is known from three different Arthropod orders;

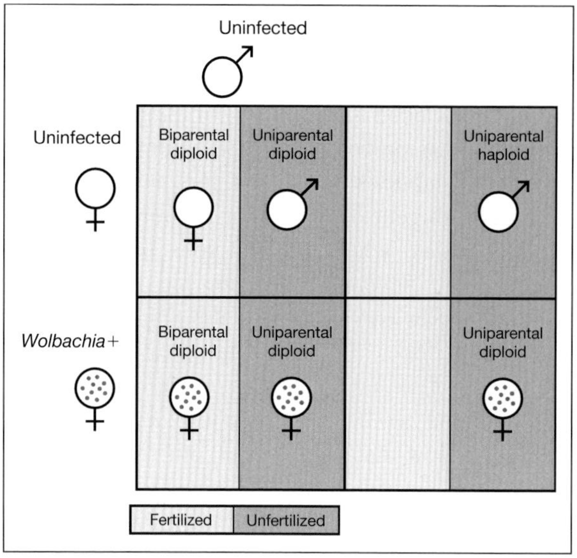

Fig. 2. Parthenogenesis. *Wolbachia* inducing parthenogenesis in a haplodiploid host. Mated uninfected females produce uninfected males and females. Unmated uninfected females produce only uninfected males. Mated infected females produce both infected diploid females from fertilized eggs and parthenogenetically derived infected diploid females. Unmated infected females produce infected diploid parthenogenic females.

Thysanoptera (Thrips), Acari (mites) and Hymenoptera (wasps). Among Thysanoptera, *Wolbachia*-induced parthenogenesis is currently known to occur in two different species, *Frankilothrips vespiformis* and *Taeniothrips inconsequens* [33, 34]. In Acari, *Wolbachia*-induced parthenogenesis is also currently known in two different species, *Bryobia praetiosa* and another undescribed *Bryobia* species [35]. Within hymenoptera however many cases of *Wolbachia*-induced parthenogenesis have been documented. These include *Aphytis lingnanensis, Aphytis diaspidis* [36], *Apoanagyrus diversicornis* [37], *Diplolepis spinosissimae* [38], *Leptopilina clavipes* [39], *Muscidifurax uniraptor* [40], and *Telenomus nawai* [41] and several species within the genus *Trichogramma* [42, 43].

It is notable that all currently documented cases of *Wolbachia*-induced parthenogenesis are found only within haplodiploid species. Haplodiploidy is a reproductive system in which females (diploid) develop from fertilized eggs, while males (haploid) develop from unfertilized eggs. There are likely several reasons for this limited distribution including; sampling bias, the mechanism(s) of sex determination in haplo-diploids, preexisting regular arrenotokous development (development of unfertilized eggs) and low levels of deleterious recessive mutations. As more *Wolbachia*-induced phenotypes are described, it will be interesting to see if the restricted distribution of parthenogenesis-inducing *Wolbachia* remains true. It should be noted that a single parthenogenic beetle, *Naupactus tesselatus*, has been reported infected with *Wolbachia* [44]. What role, if any, *Wolbachia* has in the induction of parthenogenesis in this species has yet to be investigated. *Wolbachia* has not been reported in association with any of the well-studied cases of parthenogenesis in *Drosophila*.

Parthenogenesis can be accomplished in two general ways: apomixis, where meiosis is completely or wholly suppressed, or automixis where diploidy is restored following meiosis. *Wolbachia*-induced parthenogenesis has been described as both apomictic and automictic. In *Wolbachia*-infected *Trichogramma* and *L. clavipes* anaphase is aborted during the first mitotic division, resulting in a diploid (female) nucleus [39, 45]. This is in contrast to the mechanism of thyletoky in *Trichogramma cacoeciae*, which is apomictic and does not involve *Wolbachia* [46]. In the wasp *M. uniraptor*, *Wolbachia*-induced thyletokous reproduction is also automictic, but with slightly different timing. The first mitotic division proceeds normally through anaphase and teleophase, resulting in two normal haploid (male) daughter nuclei. These nuclei then duplicate, but do not further divide, restoring diploidy, and resulting in female development [47]. In mites, within the genus *Bryobia*, *Wolbachia*-induced parthenogenesis is thought to be functionally apomictic, with heterozygosity easily detected in parthenogenetically derived diploid females. It is unclear if this is the result of a premeiotic doubling or true apomictic parthenogenesis [35]. *Wolbachia* cause parthenogenesis in at least 3 different ways (2 automictic and 1 apomictic), it is yet to be determined what similarities exist, if any, between the molecular mechanisms leading to these different modes of parthenogenesis.

As with the *Wolbachia*-induced male killing in *O. scapulalis* via lethal feminization, the distinction between the different *Wolbachia*-induced phenotypes is blurred when considering parthenogenesis induction within haplodiploids. There is not a clear distinction between feminization of genetic males and parthenogenesis in most haplodiploids. Lacking genic or chromosomal sex determination, sex is normally determined by fertilization, with fertilized eggs developing female and unfertilized as male (an exception being those with complimentary sex determination). With *Wolbachia*-induced parthenogenesis, unfertilized eggs (initially male) develop as females. This is especially so for the cases of automixic thyletoky, where females only have one set of (duplicated) chromosomes. So again, while the distinction between the four known *Wolbachia*-induced phenotypes are convenient categories, they may artificially make distinctions, which in reality are not reflected by the biology and evolutionary history of *Wolbachia* and may impede true understanding of *Wolbachia*-induced phenotypes.

Unlike with *Wolbachia*-induced male killing, feminization or CI, *Wolbachia*-induced parthenogenesis can be an evolutionary one-way path. With both male killing and feminization, *Wolbachia*-infected females rely on uninfected males (or infected males not exhibiting phenotype) each generation. With *Wolbachia*-induced parthenogenesis however, males, and those traits found solely in males, are unnecessary. Upon the acquisition of *Wolbachia*-inducing parthenogenesis, Muller's ratchet (the accumulation of mutations within genes released from selection) may eventually lead to the decay of male-specific traits. Male-specific traits may also become vestigial due to sexually antagonistic pleiotropy, selection for traits in females with a resulting negative impact on male-specific traits. This vestigialization is not restricted to males, and can occur within any trait required only for sexual reproduction (mating behavior, sperm storage, fertilization). *Wolbachia*-induced parthenogenesis can easily be cured with antibiotics, and male production restored. The resulting males however are not always fully functional. While in some species the resulting males seem to be fully functional [45], in others males lack the ability to produce sperm [48] or court females [49]. In still other cases, some sperm are produced, stored within females, but are not competent for fertilization [36]. Likewise, females in some *Wolbachia*-induced parthenogenetic species have lost the ability to correctly store sperm [37, 48, 50] or court altogether [37, 48]. The genetic basis of the loss of sexual traits due to *Wolbachia*-induced parthenogenesis has yet to be determined. *Wolbachia*-induced parthenogenesis provides excellent model systems to discriminate between Muller's ratchet and antagonistic pleiotropy in the vestigialization of traits.

In those species where parthenogenesis-inducing *Wolbachia* has not yet led to irreversible vestigialization, sexual reproduction may still occur. In some species of the wasp genus *Trichogramma*, both infected and uninfected individuals can coexist in a single population. When an uninfected male mates with and infected female, normal daughters are produced from fertilized eggs. *Wolbachia* does not interfere with fertilization and subsequent development. Parthenogenesis is only induced within unfertilized eggs, resulting in thelytokous females. When unmated, an infected female will produce thelytokous daughters [45]. In such a mixed population, males can become scarce. Consequently, uninfected females that produce increased proportions of males will theoretically be favored [51]. Such a mutation, that favors the production of males, has recently been described in the wasp *T. nawai*. Uninfected females with the male-producing genotype, produce high numbers of males, even after mating. This 'female functional virginity' is determined largely by a single locus, or multiple closely linked loci [52].

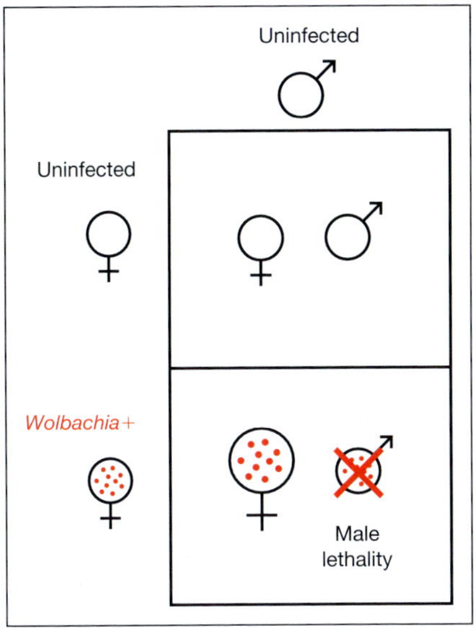

Fig. 3. Male killing. *Wolbachia*-induced male killing in a diploid host. Infected (red) females mated to uninfected males produce infected females and inviable infected males. Resource reallocation from dead males to infected female siblings increases the fitness of infected females.

Male Killing

The third phenotype caused by *Wolbachia* is the killing of genetic males (fig. 3). The advantage of male killing by *Wolbachia* may not be immediately obvious. Theory predicts that a male killing *Wolbachia* infection will only be advantageous under limited conditions, where the death of males will have a positive impact on closely related siblings [53]. Other benefits of male killing may include the resulting avoidance of inbreeding [54]. Consequently, only hosts with high sibling competition for resources are those in which male-killing *Wolbachia* should exist. To date, male-killing *Wolbachia* infections have been described in four different Arthropod orders. Within insects, these include Diptera (*Drosophila bifasciata* and *Drosophila innubila*) [55, 56], Coleoptera (*Tribolium madens* [57] and *Adalia bipunctata* [58]) and Lepidoptera (*Acraea encedon* [59] and *O. scapulalis* [32]). Outside of Insecta, male killing has been reported in pseudoscorpiones (class Arachnida) in the pseudoscorpion *Cordylochernes scorpioides* [60].

Male-killing infections are only likely to become established and spread if the death of males is beneficial to the survival and fecundity of their infected female siblings. Each of the current known cases of *Wolbachia*-induced male killing meet this criterion. The most obvious example is the viviparous pseudoscorpion *C. scorpioides*. Early death of male embryos should provide an immediate and direct reallocation of maternal resources from males to their remaining infected (female) siblings [60]. The spread of male-killing *Wolbachia* in *C. scorpioides* however may be severely limited due to higher rates of total brood abortion from infected versus uninfected mothers [61].

The male-killing *Wolbachia* infections in *D. bifasciata* and *D. innublia* both exist in an environment with high rates of sibling competition, satisfying the requirements for the persistence of a male-killing infection. *D. bifasciata* lay eggs in deposits of tree sap fluxes, which may be a limited resource [55, 62], while *D. innubila* lay eggs in mushrooms, where females typically deposit many eggs on a single mushroom. *D. innubila* lives in a very water-limited environment, where mushrooms are scarce or absent for most of the year, especially so in times of drought [56].

It is notable that male-killing *Wolbachia* has yet to be described in parasitoid hymenoptera, where either CI or parthenogenesis causing *Wolbachia* is common. Some parasitoids lay multiple eggs within a host and competition for food can occur. In these species, high sibling competition may be expected to favor male killing, and at least one other male-killing bacteria has been described [63]. This may be because the mechanism(s) employed by *Wolbachia* are incompatible with male killing in hymenoptera. As more *Wolbachia*-induced phenotypes are described, it will be interesting to see if this trend continues.

Cytoplasmic Incompatibility

The most commonly described, and phylogenetically diverse *Wolbachia*-induced phenotype is CI, currently known from at least eight different arthropod orders: Acari [64], Coleoptera [65], Diptera [66], Isopoda [67], Lepidoptera [68], Hymenoptera [69], Homoptera [70] and Orthoptera [71]. CI is manifest when a *Wolbachia*-infected male mates with a female lacking the same *Wolbachia* type (either uninfected or infected with a different *Wolbachia* type) (fig. 4). All other combinations of crosses are compatible. In diploid organisms, the result of an incompatible cross is increased embryonic mortality. At its most extreme, when a *Wolbachia*-infected male mates with an uninfected female, all offspring die (incompatible cross). The same infected male mated to a similarly infected female (compatible cross) sees no increase in offspring mortality. The spread of such *Wolbachia* through a population is easily explained theoretically. Briefly, in a mixed population (with both infected and uninfected individuals), the presence of *Wolbachia*-infected males increases the relative fitness of infected females by reducing the fitness of uninfected females. The speed at which *Wolbachia* can spread is dependent on the proportion of offspring affected

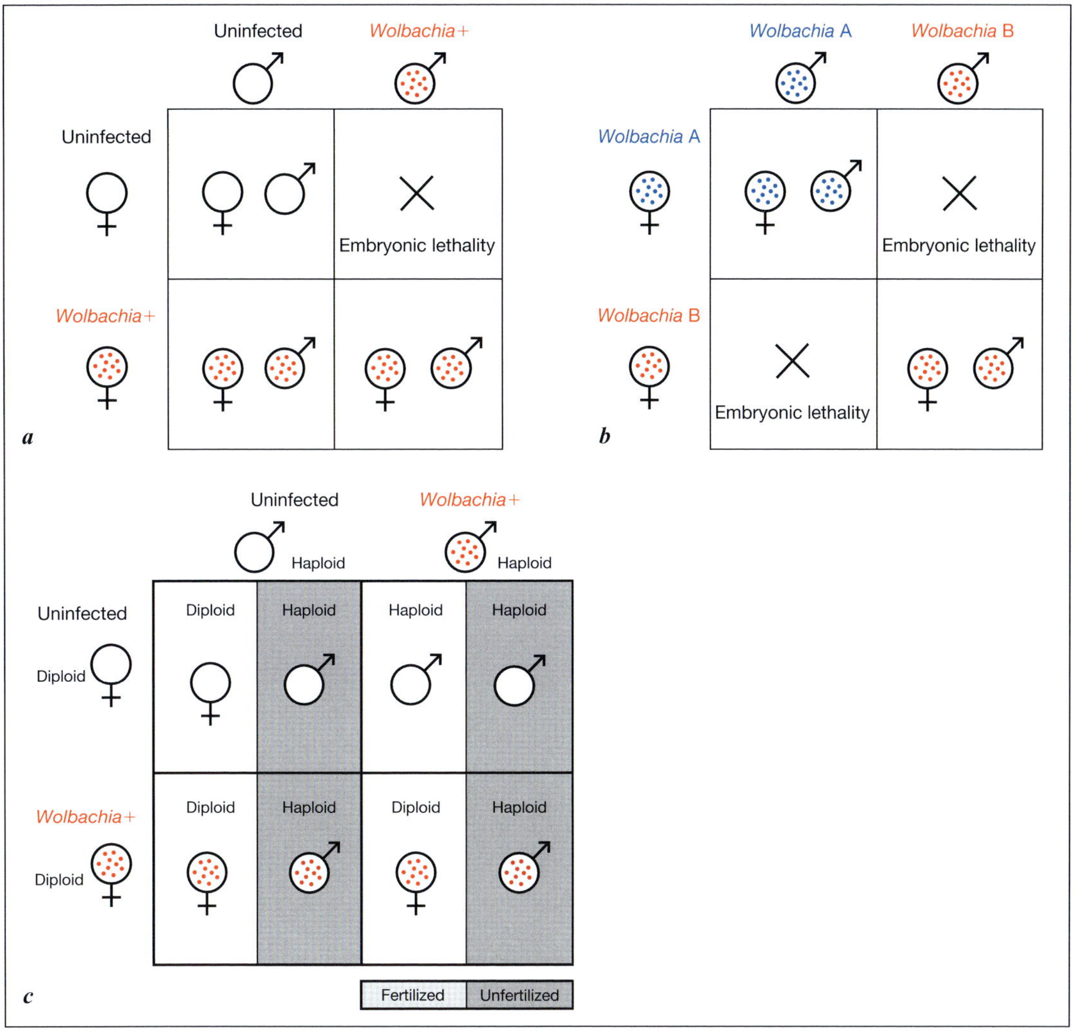

Fig. 4. ***a*** Unidirectional CI in diploids. Infected males mated to uninfected females results in early embryonic lethality. All other crosses are compatible. ***b*** Unidirectional CI in diploids. Infected males only successfully produce offspring with females with the same *Wolbachia* type. Males infected with *Wolbachia* A (blue) mated to females with *Wolbachia* B (red) results in embryonic lethality. Conversely, males with *Wolbachia* B (red) mated to females with *Wolbachia* A (blue) results in embryonic lethality. Only crosses between similarly infected individuals are compatible. ***c*** Unidirectional CI in haplodiploids. Uninfected females mated to uninfected males produce both uninfected diploid females (fertilized eggs) and uninfected haploid males (unfertilized eggs). Mated infected females (red) produce both infected diploid females (fertilized eggs) and infected haploid males (unfertilized eggs) regardless of paternal infection status. Uninfected females mated to infected males (incompatible cross) produce only uninfected haploid males from fertilized and unfertilized eggs.

in an incompatible cross, *Wolbachia* transmission rate and other effects on host fitness [72–76]. The effects of CI *Wolbachia* have been described in nature. The spread of a CI-causing *Wolbachia* was documented while spreading through previously uninfected populations of *Drosophila simulans* in California in the 1980s [74, 76, 77].

There are at least two distinct events involved in CI, the first is the modification of sperm. Since no *Wolbachia* are present in mature sperm [78, 79], *Wolbachia* presumably exert their effect prior to the conclusion of spermatogenesis. In *Drosophila*, all *Wolbachia* are removed from developing sperm late in development, at the time of spermatid individualization [79–81]. The second event is either the rescue of that modified sperm within a similarly infected egg upon fertilization, or the resulting embryonic defects and mortality upon fertilization of an uninfected egg (or an egg lacking the same *Wolbachia* type(s) found in the male). The exact nature of sperm modification or rescue is currently unknown. Some hypothesis regarding the mechanism(s) of CI will be discussed later. The following is a summary of descriptions of the embryonic defects caused by an incompatible cross.

The Cellular Biology of CI

The early embryonic defects associated with CI are very similar in each species in which it has been described. This is true for *D. simulans* [82, 83], *D. melanogaster* [84], *Culex pipiens* [85], *Nasonia vitripennis* [69, 86] and *A. vulgare* [67].

A number of well-characterized abnormalities have been documented in embryos created from incompatible crosses in *D. simulans*. Central to the incompatible phenotype is an asynchrony in the development of male and female pronuclei. At pronuclear apposition, both pronuclei are normally similar in appearance. In an incompatible cross however, the male pronucleus is more condensed than the female pronucleus [82, 83]. The following mitoses are then defective and asynchronous, beginning with the first nuclear mitosis. Chromatin defects are commonly seen, with bridges of chromatin stretching between dividing nuclei [82, 83, 87]. Centrosome duplication occurs without nuclear duplications, resulting in overproliferation of centrosomes [82, 83]. Not all embryos die early in development, with some embryos surviving until past gastrulation, but with clear nuclear and other defects [82].

In the wasp *N. vitripennis*, similar defects in the first mitotic division have been observed [69]. A cytological analysis has revealed that not only are the male and female pronuclear chromatin development asynchronous, but differences in the timing of pronuclear envelope breakdown precedes the differences in chromatin condensation. Male pronuclear envelope breakdown lags behind that of female pronuclear envelope breakdown. This suggests that *Wolbachia* may not simply be modifying sperm chromatin, but modification may involve other components of cell cycle regulation [86].

While in diploid organisms an incompatible cross results in embryonic mortality (mortality CI, fig. 4a), in haplo-diploid hosts, an incompatible cross may have a different outcome. Rather than embryonic mortality, the result can be the production of haploid (male) offspring (conversion CI, fig. 4c). Among the three closely related species of the parasitic wasp *Nasonia*, both mortality and conversion CI can be found. Conversion CI is normally found in *N. vitripennis*, while mortality CI is normally found in *N. giraulti* and *N. longicornis*. This difference is largely due to host nuclear factors and not *Wolbachia* type [88]. Mechanistically, mortality and conversion CI (dead versus male progeny) are not as different as they may first appear. Common to all CI is the delayed development of the male pronucleus relative to the female pronucleus. In mortality CI, when the first mitotic division begins, the chromosomes from the male pronuclei cannot segregate properly and are stretched and fractured to a degree that makes successful development impossible. Conversion CI can be viewed cytologically as more extreme than mortality CI, with male pronuclear development further slowed with respect to female pronuclear development. When the first mitotic division begins, the male pronucleus is entirely excluded from the process. The result is haploid male development [89]. Similar haploid embryonic development has been described in diploids as a consequence of an incompatible cross [83, 90], but haploid development in diploids is usually lethal. In *C. pipiens*, in an incompatible cross, 0.1% of embryos escape mortality and develop parthenogenically. This likely involves the complete exclusion of the male chromosomes, as in conversion CI, and possible diploidy restoration, as in parthenogenesis [85].

It is notable that several mutant phenotypes described in early embryos of *D. melanogaster* resemble those defects caused by *Wolbachia*-induced CI. These include mutants in the genes *maternal haploid, ms*(3)*K81* and *sésame.* Like CI, the phenotype resulting from mutations in these three genes each includes abnormal asynchrony in maternal and paternal pronuclei development, defective mitoses and chromatin defects. While *maternal haploid* and *sésame* are maternally expressed genes, *ms*(3)*K81* is paternally expressed [91–93]. What connection, if any, the functions of these genes have in the expression of *Wolbachia*-induced CI has yet to be determined.

CI and Speciation

It is immediately obvious how *Wolbachia*-induced CI may contribute to reproductive isolation and speciation. In the simplest case, with bidirectional CI (fig. 4b), two populations fixed for two incompatible *Wolbachia* types with complete penetrance are essentially completely reproductively isolated. Theoretical studies support a potential role of *Wolbachia* in speciation [94–97]. *Wolbachia* has been implied to contribute to reproductive isolation at hybrid zones [98], and shown to contribute to species barriers between *Drosophila recens* and *Drosophila subquinaria* [99]. In one of the most well-studied *Wolbachia*-host systems, the *Nasonia* species group, *Wolbachia* is the most significant barrier between species. *Wolbachia*-induced isolation likely preceded other reproductive barriers in this species group [100]. Theoretical studies support a role for *Wolbachia* in evolution of premating isolation and stabilizing genetic differences between diverging populations [95, 96]. Until the role of *Wolbachia* in reproductive isolation is closely examined in other host species, the exact role of *Wolbachia* in speciation will remain controversial [19, 101].

Models of CI

The exact molecular mechanism(s) of CI is currently unknown. Several different types of mechanisms have been hypothesized [102]. It should be noted that each of these models are not mutually exclusive. In one model, *Wolbachia*-derived molecules bind to some component of sperm rendering it unable to successfully initiate normal development upon fertilization. *Wolbachia* (or *Wolbachia*-derived products) in eggs then remove the sperm-bound molecule, returning normal sperm function [103]. In another proposed model, *Wolbachia* during spermatogenesis act as a sink for some vital host component of sperm. In the egg, *Wolbachia* releases this vital component, allowing for restoration of sperm function and normal development [104]. In yet another model, *Wolbachia* modify sperm so that male pronuclear development is slower than normal. The *Wolbachia* within an egg rescue development by similarly slowing down the development of the female pronucleus [86, 90]. The modification of host gene expression during spermatogenesis has also been suggested to play a role in CI. Presumably similar, or functionally complimentary, changes in egg gene expression could be involved in CI rescue. Recently, an overabundance of the mRNA and protein product of the cytoplasmic myosin heavy chain gene *zipper* has been shown to be correlated with *Wolbachia* infection during spermatogenesis in *D. simulans*. Artificial overexpression of the *zipper* gene in *D. melanogaster* males in the absence of *Wolbachia* resulted in defects in spermatogenesis and early embryonic defects upon mating with normal females [105]. As *Wolbachia*-infected females were unable to rescue this CI-like phenotype, it is unclear exactly what role cytoplasmic myosin expression may play in *Wolbachia*-induced CI.

Other *Wolbachia*-Induced Phenotypes?

Although *Wolbachia* are normally described as manipulating hosts via the four phenotypes described above, it is becoming increasingly clear that

Wolbachia may have other effects on hosts in as yet little understood ways. Care should be taken when considering other, sometimes subtle effects of *Wolbachia*. Host nuclear backgound, as well as mitochondrial type should be strictly controlled. For example, in *N. vitripennis*, initial reports suggested beneficial effects of *Wolbachia* infection on host fecundity [106]. However, in an experimental design controlling for host genetic background, no effects of *Wolbachia* on fecundity were observed [107].

Within *D. melanogaster*, *Wolbachia*-induced CI has been described numerous times [16, 81, 84, 108–112]. Some other, subtler effects of *Wolbachia* in *D. melanogaster* are now becoming apparent. The first such interaction was uncovered when stocks carrying an allele of the sex-determining gene *Sex lethal* (*Sxl*) was found to be infected with *Wolbachia*. The expression of the *Sxl* mutant phenotype was in an unusual, non-Mendelian manner. Females homozygous for Sxl^{f4} are normally sterile. Ovaries from homozygous Sxl^{f4} females normally are much smaller than normal, with few or no late-stage eggs. However, when a new stock with homozygous Sxl^{f4} females was created in preparation for a suppressor screen, females were already weakly fertile prior to the beginning of the screen. The ovaries from homozygous Sxl^{f4} females in the new stock had large numbers of late-stage eggs. It was later determined that the suppressor of Sxl^{f4} in this stock was not paternally inherited, or even nuclear, but rather cytoplasmic. The stock was then determined to have been *Wolbachia* infected. Subsequent tetracycline treatment eliminated the Sxl^{f4} suppressor, confirming a likely role of *Wolbachia* in Sxl^{f4} suppression. *Wolbachia* also had a similar but reduced effect on some, but not all, other *Sxl* alleles tested. The nature of the interaction between *Wolbachia* and *Sxl* is still unclear, but likely does not simply involve the bypass of *Sxl* or increased expression [113]. Although interactions between *Wolbachia* and sex determination have been implicated in a number of different systems [32, 114, 115], it is highly improbable that an interaction with *Sxl* is a unifying principal in *Wolbachia*'s manipulation of host reproduction. Although *Sxl* is a master switch in sex determination in *D. melanogaster*, it is not involved in sex determination, or even sex specifically regulated outside of the genus *Drosophila* [116, 117].

Another unusual and little understood effect of *Wolbachia* within *D. melanogaster* was found in two stocks containing mutants in the insulin receptor substrate gene, *chico*. Flies homozygous for mutations in *chico* are markedly smaller than normal. While investigating the effects of *chico* mutants in work unrelated to *Wolbachia*, it was determined that two stocks containing *chico* mutations were infected with *Wolbachia*. To avoid any potentially confounding effects of *Wolbachia*, these stocks were treated with tetracycline. Following the removal of *Wolbachia*, the recovery of mutant *chico* homozygotes was greatly reduced in both stocks. In one of the stocks, no homozygotes were ever recovered in the absence of *Wolbachia*. When *Wolbachia* was reintroduced into these stocks via embryonic microinjection, homozygotes were again recovered. The exact nature of the effect of *Wolbachia* in this system is unclear, as the effect is absent in different nuclear backgrounds [84]. With a high proportion of *D. melanogaster* stocks used in research infected with *Wolbachia* [84, 110], it will be interesting to see how many other such interactions are described in the future.

Aside from interactions with mutant phenotypic expression in *Drosophila* research, *Wolbachia* is interacting with hosts during oogenesis in as yet little understood ways. A naturally occurring obligate association with *Wolbachia* has been described in the wasp *Asobara tabida*. Normally infected, removal of *Wolbachia* prevents the completion of oogenesis. *Wolbachia*-free females fail to produce mature eggs, and consequently fail to reproduce. All other aspects of *A. tabida* biology, including spermatogenesis, seem unaffected by the removal of *Wolbachia* [118]. *A. tabida* is normally coinfected with three distinct *Wolbachia* strains. It is however only one of the three that is obligatory for successful oogenesis. The other two *Wolbachia* types are able to induce CI [119]. Not only is this system remarkably similar to that seen with *Sxl* in *Drosophila* [113], it is also similar to effects seen with *Wolbachia* infections in filarial nematodes [120]. Recently, another apparent case of symbiont-dependent oogenesis has been described. The date stone beetle, *Coccotrypes dactyliperda*, is infected with both *Wolbachia* and *Rickettsia*. Removal of bacteria with antibiotics significantly reduces egg production. Which bacteria (*Wolbachia* or *Rickettsia*) are necessary for oogenesis is currently unclear [121].

As described previously, the nature of the maternal inheritance of *Wolbachia* can result in selection for infected female hosts over males. As a result, parthenogenesis, feminization and male killing can bias the proportion of *Wolbachia*-infected females in a population. In at least one system however, the proportion of females is increased as a result of *Wolbachia* infection, but not via the usually described phenotypes. In the two-spotted spider mite, *Tetranychus urticae*, significantly more females are produced from infected versus cured mothers. Feminization, parthenogenesis, male killing as well as CI have been eliminated as possible explanations for this *Wolbachia*-induced phenotype. As sex in this species is determined by the proportion of eggs fertilized (males develop as haploids from unfertilized eggs), it is likely that *Wolbachia* is somehow interacting at this level. Over several generations, lines cured of *Wolbachia* eventually approach the sex ratio of infected lines. This suggests that *Wolbachia* and host are engaged in an arms race to control sex ratio, with *Wolbachia* favoring a sex ratio more female biased than is optimal for the host. This struggle for the control of sex ratio is likely at a dynamic equilibrium in infected lines, at or near the sex ratio optimal for the host. The removal of *Wolbachia* results in a sex ratio more male biased than is optimal [122].

In addition to the direct manipulation of reproduction, an endosymbiont may increase the prevalence within a population by providing positive fitness benefits. Several studies have searched for such benefits associated with *Wolbachia* with mixed results. There has been considerable interest in potential fitness costs or benefits of *Wolbachia* in *D. melanogaster*. In *D. melanogaster*, rates of CI are usually low or nonexistent under laboratory conditions, and likely lower in the field [123]. Consequently, other benefits of *Wolbachia* infection must be invoked to explain the persistence of *Wolbachia* within *D. melanogaster* in natural populations. A recent study has suggested that one of five identified *Wolbachia* variants found within *D. melanogaster* has swept though global populations within the past century [124], adding to the inadequacy of current theory (based on CI) in explaining the spread and persistence of *Wolbachia* within this species. Additional mutualistic interactions must be invoked to explain the behavior of *Wolbachia* in *D. melanogaster*. In one study, the costs and benefits of *Wolbachia* infection in *D. melanogaster* varied among different fly strains. Different cured fly lines showed increased, decreased or unchanged longevity, depending on host genotype [125]. The effects on survival and fecundity are more consistent. Females from three of four lines showed a significant reduction in survival and fecundity (egg laying) following the removal of *Wolbachia*. With one exception, there was no significant change in egg to adult viability following *Wolbachia* removal. The one exception was reduced viability when uninfected females were mated to uninfected males as compared to infected females mated to uninfected males [126].

Other studies suggest a potential role of *Wolbachia* in host aging. Isonuclear lines of *D. melanogaster*, differing only in cytoplasm showed differences in longevity [127]. These data were initially interpreted as support for the pivotal role of mitochondria in aging. After tetracycline treatment however, the *Wolbachia*-free lines no longer showed different rates of longevity [128]. *Wolbachia* then can potentially contribute to differences in host longevity. Significant effects of *Wolbachia* on host longevity in *D. melanogaster* have also been well documented due to the *Wolbachia* variant *w*MelPop (popcorn). This interaction was unexpectedly uncovered in a *Drosophila* mutagenesis screen. Flies with *w*MelPop suffer significant reduction in longevity. This is likely due to the overproliferation of *Wolbachia*, especially in neuronal tissue [129].

As the degree to which *Wolbachia* can affect a host has been shown to differ under ideal laboratory and more stressful field conditions [123], a search for possible fitness benefits from *Wolbachia* in *D. melanogaster* was undertaken under stressful conditions. No benefits of *Wolbachia* were measured in adult starvation resistance, or larval development time and adult size under nutrient-limiting conditions. There was also no effect of *Wolbachia* on heat resistance, survival or virility under heat-stressed conditions [130].

Within *Aedes albopictus*, fitness benefits resulting from *Wolbachia* infection (aside from CI) are far more obvious than in *Drosophila*. *Wolbachia* infection confers both a significant fecundity and longevity benefit. Both singly and doubly infected females produce more eggs than uninfected females within the same host genetic background. In addition, both singly and doubly infected females lived longer than uninfected. There is no effect of *Wolbachia* on male longevity [131]. The exact nature of the fitness benefit conferred with *Wolbachia* has yet to be determined.

The potential effect of CI causing *Wolbachia* on reproductive isolation and speciation is a topic that has received considerable attention and is discussed extensively elsewhere [75, 94, 95, 97, 100, 132]. A recent study with *Drosophila* however has provided direct evidence that *Wolbachia* may be capable of influencing sexual isolation between populations aside from CI. Different laboratory populations of *D. melanogaster* with a common origin underwent selection over several decades for tolerance to toxins in food. One group of populations was selected for heavy metal tolerance, while another was selected for ethanol tolerance. Five of eight of the resulting populations were later determined to be infected with *Wolbachia*. Some of these populations now show high levels of behavioral isolation manifest in mate discrimination. Removal of *Wolbachia* greatly reduces the degree of mate discrimination [133]. Thus *Wolbachia* has the potential to be an important factor in both pre- and postzygotic isolation between populations.

Although *Wolbachia* successfully evade the host immune system, and do not induce the normal antibacterial response [134], *Wolbachia* infection has been shown to be an important factor in host immunity. In at least one host-parasitoid system, the presence of *Wolbachia* decreases fitness in both the host and parasitoid. *D. simulans* infected with *Wolbachia* were less able to encapsulate and kill eggs laid by the parasitoid *Leptopilina heterotoma*. Similarly, in the parasitoid *L. heterotoma, Wolbachia* removal is beneficial, with the uninfected parasitoid more likely to evade the host defenses. In this system then, *Wolbachia* in the host makes it more susceptible to parasitoids, while *Wolbachia* in the parasitoid, makes it more likely to be detected and killed by the host [135]. The exact nature of these interactions is currently unknown.

What effect *Wolbachia* has on sperm competition is currently unclear. Wade and Chang [136] are often cited as evidence for a positive impact of *Wolbachia* infection on sperm competition in *Tribolium confusum*. Females given different numbers of uninfected and infected mates had offspring fathered by a disproportionate number of infected males. As this study did not record mating order or frequency, the exact impact of *Wolbachia* on sperm competition is unclear and indirect at best. Another study often cited when referring to a potential negative impact on sperm production and sperm competition is Snook et al. [79]. This study has been cited as showing that infected males produce fewer sperm than uninfected males. More accurately however, this study concludes that the testes of infected males of one specific age have fewer sperm cysts of one specific developmental stage. This may reflect differences (either increased or decreased) in sperm production or the rate of sperm production, but the study did not distinguish between the two. Total sperm production was not addressed. Hoffmann et al. [74] directly tested sperm competition and *Wolbachia* infection using virgin males in *D. simulans*. They concluded that *Wolbachia* had no effect on sperm competition. Recently however Champion de Crespigny and Wedell [137] investigated sperm competition and *Wolbachia* in *D. simulans* using nonvirgin males and found a significant effect. In the second male role, nonvirgin infected males sired fewer (~10% less) offspring than uninfected males. It is unclear if this reduction is due to a reduction in sperm production with *Wolbachia* infection, or reduced competitive ability of sperm from *Wolbachia*-infected males.

Novelty and Ancestry?

Wolbachia are not novel in the ability to manipulate the reproduction of hosts. Similar selection pressures coupled with an intracellular lifestyle are expected to favor reproductive parasitism in any maternally inherited microorganism [13]. As expected, other microorganisms have evolved the ability to induce feminization, parthenogenesis, male killing and CI, and the scope of microbes performing these effects is just coming to light. However, so far, *Wolbachia* has the greatest spectrum of host reproductive manipulations of any bacteria group.

Feminization

Feminizing microorganisms have been reported a number of times. *Nosema granulosis* is a microsporidian parasite found in several isopods including *Gammarus duebeni, Orchestia aestuarensis, Orchestia gammarellus* and *Orchestia mediterranea*, where it converts genetic males into functional females [138, 139]. Also like *Wolbachia*, this feminization in isopods is accomplished through disruption of the androgenic gland and resulting androgenic hormone [140].

Feminization caused by Cytophaga Flexibacter Bacteroides (now referred to as *Cardinium*) has been described within the false spider mite, *Brevipalpus phoenicis*. This species normally consists of only females. Haploid eggs develop as females, not through diploidization, but through true feminization [141].

Parthenogenesis

At least two different cases of bacteria, other than *Wolbachia*, causing parthenogenesis have been reported. Recently, a *Rickettsia* was shown to induce thylytokous parthenogenesis in the wasp, *Neochrysocharis formosa*, an endoparasite of leaf miners [142]. In addition, *Cardinium* infection (previously referred to as *Cytophaga*-like organism) within two species of *Encarsia*, the wasp parasitoid of the Whitefly, is associated with parthenogenesis. Within *Encarsia hispidia, Cardinium* has been confirmed to be the causative agent of parthenogenesis [143, 144].

Male Killing

The reproductive phenotype reported associated with the highest diversity of microorganisms is male killing. Several different microorganisms have now been shown to induce male killing. *Spiroplasma* bacteria have been described as male killers in a wide variety of hosts. These include several *Drosophila* species (Diptera) [145, 146], as well as Coleoptera, *A. bipunctata, Anisosticta novemdecimpunctata* and *Harmonia axyridis* [147–149], and Lepidoptera, *Danaus chrysippus* [150].

Male-killing *Rickettsia* have been described in three different beetle species, the ladybird beetles *A. bipunctata, Adalia decempunctata* and the leaf mining beetle *Brachys tessellates* (Coleoptera) [151–153]. Ladybird beetles are particularly susceptible to invasion by male-killing microbes due to the benefit of early resource reallocation. Single females will typically deposit eggs in clutches, which soon after hatching require a meal for survival. When males are killed as embryos, the surviving infected female siblings feed on their dead brothers. This reallocation of resources from males to females provides an immediate and significant benefit to females, greatly increasing the rate of survival compared to females with living brothers [154].

In addition to *Wolbachia, Spiroplasma* and *Rickettsia* infections, another bacterium causes male killing in ladybird beetles. *Adonia variegata* and *Coleomigilla maculata* have been shown to be infected with a male-killing flavobacterium [155, 156].

Another bacterium, *Arsenophonus nasoniae*, is a specific male killer. Originally described in *N. vitripennis* (Hymenoptera), *A. nasoniae* kills only arrenotokously created males (from unfertilized eggs). Males arising from fertilized eggs (produced through *Wolbachia*-induced CI, or the selfish genetic element PSR) are not susceptible to this male killing [63].

Still other male killing microorganisms have yet to be identified. For example, in some strains of the oriental tea tortrix, *Hypolimnas bolina* (Lepidoptera), all female broods are observed. In such broods, mortality is greather than 50%, suggesting the sex ratio bias is due to male killing. The male killing trait was maternally inherited, but resistant to antibiotic treatments known to cure *Wolbachia, Rickettsia* or *Spiroplasma*. The agent of male killing was easily transmitted by feeding the remains of dead male larva from broods with sex ratio bias to feeding larva from non-sex ratio-biased lines. This excludes nuclear factors or mitochondria as the male killing agent [157].

Cytoplasmic Incompatibility

Although the most widely described *Wolbachia*-induced phenotype is CI, other CI-causing bacteria are little known. Only recently has another bacteria been shown to induce CI. Found in the parasitic wasp, *Encarsia pergandiella*, this is the third reproductive phenotype known caused by *Cardinium*. When infected males are mated to uninfected females, the result is female embryonic mortality [158]. These symbionts are closely related to the bacteria responsible for parthenogenesis in *E. pergandiella* [143, 158] and feminization in *B. phoenicis* [159]. A recent survey has revealed that ~7% of arthropods tested contained these bacteria [159] (alternately referred to as *Encarsia* bacterium [143], *Cytophaga*-like organism [159] or *Candidatus Cardinium hertigii* [144]).

As work continues on the exact molecular mechanisms behind microorganism-induced reproductive phenotypes, it will be interesting to determine what similarities, if any, exist between the mechanism employed by *Wolbachia* and other microorganisms.

Can the phylogenic relationships among the above-mentioned bacteria and *Wolbachia* suggest anything about the origin and evolution of their induced phenotypes? The closest bacteria to *Wolbachia* known to manipulate reproduction of invertebrate hosts are the *Rickettsia*. Multiple *Rickettsia* are known to cause male killing, while one *Rickettsia* causes parthenogenesis. Based on these data alone then, parsimony may suggest that male killing is the ancestral phenotype of *Wolbachia*. This conclusion is likely oversimplified. As CI is by far the most widespread *Wolbachia*-induced phenotype found in the widest range of hosts, it is also a good candidate for an ancestral phenotype. Until the exact molecular mechanisms underlying the different reproductive phenotypes are determined, any attempts to determine ancestral phenotype will likely be largely speculative.

Wolbachia-Host Interactions

The first appearance of *Wolbachia* in the scientific literature [4] was a description of bacteria within the ovaries of mosquitoes. Since then, a number of studies have described *Wolbachia* and their interactions with hosts at the cellular level. The following sections will review the current knowledge on *Wolbachia*-host interactions.

Spermatogenesis

Likely the first cytological description of *Wolbachia* (although not then identified) within *Drosophila* was during spermatogenesis in *D. melanogaster* [160]. To date, the most detailed description of *Wolbachia* during spermatogenesis comes from studies in *D. simulans*. The following section describes the behavior of *Wolbachia* during spermatogenesis in *D. simulans*.

Spermatogenesis begins with the spermatogonial stem cells at the apical hub of the testis. Although *Wolbachia* have not been specifically described in the spermatogonial stem cells, its presence is inferred from descriptions of *Wolbachia* in earlier (pole cells and later embryonic germ cells) [87, 161] and later (mitotically dividing spermatogonial cells) [81, 162] within the germ line. In *Drosophila*, as in other arthropods, sperm mature within a cyst comprised of a germline (gonial cells, spermatocytes and later spermatids) and somatic (cyst cells) component. A single primary gonial cell is created as the daughter cell of a spermatogonial stem cell, while two cysts cells are the product of somatic cyst progenitor cells. Each stem cell is located at the apical end of the testis within the stem cell niche [163]. A cyst is created when a primary gonial cell (germline) is surrounded by two cyst cells (somatic). Both the germline as well as somatic component of a cyst may contain *Wolbachia*, presumably arising from the respective stem cells [164]. There is evidence to suggest that *Wolbachia* in *Drosophila* spermatogonial stem cells either do not replicate, or do not do so at a sufficient replacement rate. As a result, the spermatogonial stem cells eventually are depleted of *Wolbachia*, and subsequently give rise only to uninfected daughter cells. Consequently, the number of *Wolbachia*-infected cysts decreases with male age. The timing of the depletion of *Wolbachia* from the male germ line is a key determinant in the rate of CI in *Drosophila* [164].

As a cyst develops, the germline undergoes four rounds of mitosis followed by meiosis (the exact number of mitotic divisions varies among species). Normally in *D. simulans*, when a cyst is infected, each of the mitotic products, the 16 spermatocytes contain *Wolbachia*. During the four rounds of mitosis however, *Wolbachia* either do not replicate, or do not do so at a rate so as to always populate each of the mitotic products (16 primary spermatocytes). When *Wolbachia* is nearly depleted from spermatogonial stem cells, the daughter cell (primary gonial cell) is likely insufficiently provisioned with *Wolbachia* so that less than 16 spermatocytes within a resulting cyst are infected with *Wolbachia*. The pattern of infection is often seen in multiples of four neighboring spermatocytes, suggesting early mitotic divisions reduced *Wolbachia* to zero in one or another daughter cell [81]. Following mitosis in *Drosophila*, the 16 primary spermatocytes enter a prolonged growth phase in which their volume greatly increases. During this stage, *Wolbachia* rapidly proliferate, with abundant *Wolbachia* throughout the cytoplasm at the time of entry into meiosis. *Wolbachia* are uniformly distributed around the cytoplasm during mitotic interphase as well as meiosis I interphase [81]. At teleophase I, a minority of *Wolbachia* are associated with the meiotic spindle poles, with most of the *Wolbachia* (like mitochondria) associated with the spindle microtubules [162]. Following meiosis, just prior to spermatid elongation, most bacteria are localized near the mitochondrial derivatives and not the nuclei [162]. At the onset of spermatid elongation, most of the *Wolbachia* is displaced away from the nuclear end of the spermatid in *D. simulans* [81]. Within infected elongating spermatids, *Wolbachia* can be found along entire length of spermatids, but not evenly distributed. The highest concentration of *Wolbachia* can be seen at the distal (tail) end of spermatids, with a second, smaller region of *Wolbachia* accumulation toward the apical end, adjacent to the nuclear bundle [81]. The diameter of the spermatids in this heavily infected distal end of infected cysts can be increased to four times that of a normal spermatid [162]. Within *D. simulans*, the distribution of *Wolbachia* within an infected spermatid is similar despite *Wolbachia* type, incompatibility type, or even in infections not resulting in CI. Therefore, incompatibility type is not simply due to differential bacterial density or location with developing sperm [81, 164]. Unlike *D. simulans*, the *Wolbachia* found within spermatids within *D. melanogaster* are differently localized. The distal end lacks the consistent heavy localization seen in *D. simulans*. Instead, *Wolbachia* are seen scattered along the length of the cyst in seemingly random localized patches [81, 164]. It is unclear what effect, if any, the location of *Wolbachia* along a developing sperm has on sperm modification and resulting incompatibility.

Electron microscopy has revealed that during spermatogenesis *Wolbachia* have dispersed chromatin and many apparent ribosomes. Like in embryos [165], *Wolbachia's* two membranes are surrounded by a third (presumably host) membrane. In early stages of spermatogenesis (mitotic and premeiotic) the area within the third membrane is minimal. During the spermatid elongation however, the space within the third membrane is greatly increased, suggesting bacterial secretion [162].

Throughout spermatogenesis, despite large quantities of *Wolbachia*, neither the morphology nor distribution of intracellular organelles is noticeably changed by the presence of *Wolbachia* as evidenced by either electron or confocal microscopy [81, 162, 164]. While the exact nature of the modification which *Wolbachia* leave on sperm is currently unknown, it is remarkable that *Wolbachia*-infected sperm cysts can give rise to fully functional sperm. Despite the large quantities of bacteria filling the cytoplasm during spermatogenesis, the resulting sperm not only function, but aside from incompatibility, seem to suffer no other obvious defects.

Wolbachia in Oogenesis and Embryogenesis

Oogenesis

The first report of *Wolbachia pipientis* came from a cytological description of the ovaries of *C. pipiens*, for which the bacteria species is now named [4]. Within *Drosophila*, probably the first cytological account of *Wolbachia* in eggs was described by Wolstenholme as dense bodies found in the cytoplasm [166]. More recent descriptions of *Wolbachia* during oogenesis and embryogenesis in *Drosophila* using confocal and electron microscopy have begun to provide insight into specific *Wolbachia* and host interactions.

During oogenesis, *Wolbachia* can be present in all cells within an ovary [165]. A comparison of *Wolbachia* distribution during oogenesis and embryogenesis among several different host-*Wolbachia* combinations has highlighted the diversity of host-symbiont interactions, even within one host species [161].

In *Drosophila*, at stage 10 of oogenesis, a developing egg is composed of a cyst of 15 interconnected polyploidy nurse cells and a single oocyte which are surrounded by a sheath of follicle cells. In *D. simulans* ovaries infected with the bacterial type *w*Ri, *Wolbachia* is largely restricted to the follicle cells. While *D. simulans* infected with the *w*No bacterial type, *Wolbachia* are found within follicle cells (although to a lesser degree than in a *w*Ri infection), but also within nurse cells and the oocyte, with a high area of accumulation in the anterior region of the oocyte. In *D. melanogaster* infected with *w*Mel, *Wolbachia* can be found within follicle cells, nurse cells, as well as within the oocyte, with a higher accumulation within the posterior region of the oocyte [161]. This highlights the diversity of *Wolbachia* tropism within closely related host species and may suggest that the different patterns of host-*Wolbachia* interactions are crucial to the manifestation of different forms of CI.

During oogenesis in *D. melanogaster, Wolbachia* can be seen both in stem cells and within the cytoplasm of the early 16 cell cysts [167]. Within follicular and nurse cells of *D. melanogaster, Wolbachia* are in close proximity to the rough endoplasmic reticulum [168]. Following cytoplasmic dumping, *Wolbachia* from the nurse cells are transferred to the oocyte. Ferree et al. [167] describe a specific stage during oogenesis within *Drosophila* ovaries (stage 12 egg chamber) in which *Wolbachia* are organized between the anterior cortex and the nucleus of the oocyte, forming a crescent of *Wolbachia* which appears to make direct contact with the nucleus. This has been observed in *D. melanogaster* (*w*Mel and *w*Melpop) as well as *D. simulans* (*w*Ri) and suggests a potential time at which *Wolbachia* may be modifying the oocyte nucleus. The exact nature of the changes in oogenesis due to *Wolbachia* is currently unknown.

Embryogenesis

Prior to fertilization in *D. simulans*, most of the *Wolbachia* (*w*Ri) within an egg are distributed evenly throughout the cortex of the egg with fewer *Wolbachia* within the interior of the egg [16, 104, 169]. With fertilization and blastoderm formation, *Wolbachia* increasingly concentrate in the embryo cortex as the nuclei migrate to the cortex [104, 169]. *Wolbachia* become associated with poles of the centrosome-organized microtubules at mitotic spindle poles, but not with spindle microtubules [104, 169]. *Wolbachia* seem to be bound to microtubules by short electron-rich bridges [169]. The accumulation of *Wolbachia* towards centrosomes functions to evenly distribute *Wolbachia* to daughter nuclei with each nuclear division. With blastoderm formation, most *Wolbachia* become associated with the apical side of the nuclei (between the nuclei and the outside of the embryo), which corresponds to the location of the centrosome. Throughout embryogenesis, *Wolbachia* do not appear to be replicating [104].

The distribution of *w*Ri *Wolbachia* within *D. simulans* embryos is only one of three distinct distributions found in *Drosophila* embryos. In addition to the cortical localization as seen with *w*Ri [104], both posterior and anterior enrichment of *Wolbachia* within eggs and embryos has been described [161]. Preferentially locating to the posterior portion of an egg is an obviously advantageous strategy for a maternally transmitted bacterium, as this is the site of pole cell formation, which becomes the germ line. A similar distribution has been reported for *Wolbachia* in the wasps *Nasonia* [170], *Trichogramma* [40], and *Aphytis* [171]. This trait either involves *Wolbachia*'s affinity for a conserved aspect of insect early embryonic patterning, or convergent evolution of pole plasm *Wolbachia* localization in multiple host lineages. By looking at natural *Wolbachia* infections as well as artificial transinfections, it was determined that the overall distribution of *Wolbachia* within a developing *Drosophila* embryo (cortical, posterior or anterior localization) was largely dependent on *Wolbachia* and not host phylogeny. While posterior or cortical localization are understandable strategies (posterior localization leads to a disproportionate numbers of *Wolbachia* within the germ line, while cortical distribution assures an even distribution throughout the organism including the germline), the anterior embryonic localization is more difficult to understand. This strategy assures that the pole cells (and resulting ovaries and testes) begin with far fewer *Wolbachia* than are found elsewhere in the embryo [161].

Scanning electron microscopy of *Wolbachia* found in *D. melanogaster* has uncovered previously unappreciated diversity in *Wolbachia* morphology. Aside from the 'standard' infection associating with microtubules and following dividing nuclei, a second morphotype not associated with microtubules but rather associated with endoplasmic reticulum has been described. In this case, the outer (third) membrane is continuous with the endoplasmic reticulum. In addition, these *Wolbachia* often lacked the disperse chromatin normally seen, as well as contained very few ribosomes [168].

Cytological descriptions alone cannot elucidate the molecular mechanisms behind *Wolbachia's* manipulation of host reproduction. They have however provided what will likely be valuable clues into the specific host-symbiont interactions. Additional work remains to determine the exact mechanisms.

Conclusions

Research on *Wolbachia* symbiosis in Arthropods began in 1924 with its discovery in mosquitoes [4]. *Wolbachia* was then largely ignored until its importance in CI was recognized in 1973 [66]. With the advent of molecular-based methods of detection and identification, the pervasiveness of *Wolbachia* in nature has begun to be fully appreciated. The past two decades have witnessed an explosion in knowledge concerning the widespread distribution of *Wolbachia* and the diversity of effects on hosts. An increasing number of researchers from a wide range of disciplines have begun to focus on *Wolbachia*. As yet, the exact nature of *Wolbachia's* manipulation of hosts remains elusive. A number of different research programs are currently addressing these questions, with a variety of approaches. The next few years will likely witness a great number of studies uncovering the exact nature of these *Wolbachia*-host interactions, and finally reveal the specific molecular mechanisms underlying *Wolbachia's* ability to manipulate their arthropod hosts.

References

1 Werren JH, Windsor DM: *Wolbachia* infection frequencies in insects: evidence of a global equilibrium? Proc R Soc Lond B Biol Sci 2000;267:1277–1285.

2 Bandi C, et al: Phylogeny of *Wolbachia* in filarial nematodes. Proc R Soc Lond B Biol Sci 1998;265:2407–2413.

3 Dumler JS, et al: Reorganization of genera in the families Rickettsiaceae and Anaplasmataceae in the order Rickettsiales: unification of some species of *Ehrlichia* with *Anaplasma, Cowdria* with *Ehrlichia* and *Ehrlichia* with *Neorickettsia*, descriptions of six new species combinations and designation of *Ehrlichia equi* and 'HGE agent' as subjective synonyms of *Ehrlichia phagocytophila*. Int J Syst Evol Microbiol 2001;51:2145–2165.

4 Hertig M, Wolbach SB: Studies on *Rickettsia*-like micro-organisms in insects. J Med Res 1924;44: 329–374.

5 Hertig M: The *Rickettsia, Wolbachia pipientis* (gen. et sp. n.) and associated inclusions of the mosquito, *Culex pipiens*. Parasitology 1936;28:453–486.

6 Werren JH, Windsor DM, Guo L: Distribution of *Wolbachia* among neotropical arthropods. Proc R Soc Lond B Biol Sci 1995;262:197–204.

7 West SA, et al: *Wolbachia* in two insect host-parasitoid communities. Mol Ecol 1998;7: 1457–1465.

8 Jeyaprakash A, Hoy MA: Long PCR improves *Wolbachia* DNA amplification: wsp sequences found in 76% of sixty-three arthropod species. Insect Mol Biol 2000;9:393–405.
9 Kyei-Poku GK, et al: On the ubiquity and phylogeny of *Wolbachia* in lice. Mol Ecol 2005;14: 285–294.
10 Rasgon JL, Scott TW: An initial survey for *Wolbachia* (Rickettsiales: Rickettsiaceae) infections in selected California mosquitoes (Diptera: Culicidae). J Med Entomol 2004;41:255–257.
11 Ricci I, et al: Searching for *Wolbachia* (Rickettsiales: Rickettsiaceae) in mosquitoes (Diptera: Culicidae): large polymerase chain reaction survey and new identifications. J Med Entomol 2002;39:562–567.
12 Kittayapong P, et al: Distribution and diversity of *Wolbachia* infections in Southeast Asian mosquitoes (Diptera: Culicidae). J Med Entomol 2000;37:340–345.
13 Werren JH, O'Neil S: The evolution of heritable symbionts; in O'Neil S, Hoffmann AA, Werren JH (eds): Influential Passengers: Inherited Microorganisms and Arthropod Reproduction. Oxford University Press, New York, 1997.
14 Wilson MJ, Weightman AJ, Wade WG: Applications of molecular ecology in the characterization of uncultured microorganisms associated with human disease. Rev Med Microbiol 1997;8:91–101.
15 Periotti MA, et al: The sex ratio distortion in the human head louse is conserved over time. BMC Genet 2004;5:10.
16 Boyle L, et al: Interspecific and intraspecific horizontal transfer of *Wolbachia* in *Drosophila*. Science 1993;260:1796–1799.
17 Kang L, et al: Superinfection of *Laodelphax striatellus* with *Wolbachia* from *Drosophila simulans*. Heredity 2003;90:71–76.
18 Frydman HM, et al: Somatic stem cell niche tropism in *Wolbachia*. Nature 2006;441:509–512.
19 Weeks AR, Reynolds KT, Hoffmann AA: *Wolbachia* dynamics and host effects: what has (and has not) been demonstrated? Trends Ecol Evol 2002;17:257–262.
20 Martin G, Juchault P, Legrand JJ: Mise en évidence d'un micro-organisme intracytoplasmique symbiote de l'Oniscoïde *Armadillidium vulgare* L., dont la présence accompagne l'intersexualité ou la féminisation totale des mâles génétiques de la lignée thélygène. CR Acad Sci Paris Ser III 1973;276:2313–2316.
21 Rousset F, et al: *Wolbachia* endosymbionts responsible for various alterations of sexuality in arthropods. Proc R Soc Lond B Biol Sci 1992;250:91–98.
22 Michel-Salzat A, Cordaux R, Bouchon D: *Wolbachia* diversity in the *Porcellionides pruinosus* complex of species (Crustacea: Oniscidea): evidence for host-dependent patterns of infection. Heredity 2001;87(pt 4):428–434.
23 Rigaud T, et al: Mitochondrial DNA polymorphism, sex ratio distorters and population genetics in the isopod *Armadillidium vulgare*. Genetics 1999;152:1669–1677.
24 Cordaux R, et al: Evidence for a new feminizing *Wolbachia* strain in the isopod *Armadillidium vulgare*: evolutionary implications. Heredity 2004;93:78–84.
25 Vandekerckhove TTM, et al: Evolutionary trends in feminization and intersexuality in woodlice (Crustacea, Isopoda) infected with *Wolbachia pipientis* (alpha-Proteobacteria). Belg J Zool 2003;133:61–69.
26 Azzouna A, Greve P, Martin G: Sexual differentiation traits in functional males with female genital apertures (male symbol fga) in the woodlice *Armadillidium vulgare* Latr. (Isopoda, Crustacea). Gen Comp Endocrinol 2004;138:42–49.
27 Moreau J, et al: Sexual selection in an isopod with *Wolbachia*-induced sex reversal: males prefer real females. J Evol Biol 2001;14:388–394.
28 Moreau J, Rigaud T: Variable male potential rate of reproduction: high male mating capacity as an adaptation to parasite-induced excess of females? Proc R Soc Lond B Biol Sci 2003;270:1535–1540.
29 Bouchon D, Rigaud T, Juchault P: Evidence for widespread *Wolbachia* infection in isopod crustaceans: molecular identification and host feminization. Proc R Soc Lond B Biol Sci 1998;265: 1081–1090.
30 Hiroki M, et al: Feminization of genetic males by a symbiotic bacterium in a butterfly, *Eurema hecabe* (Lepidoptera: Pieridae). Naturwissenschaften 2002;89:167–170.
31 Kageyama D, et al: Feminizing *Wolbachia* in an insect, *Ostrinia furnacalis* (Lepidoptera: Crambidae). Heredity 2002;88:444–449.

32 Kageyama D, Traut W: Opposite sex-specific effects of *Wolbachia* and interference with the sex determination of its host *Ostrinia scapulalis*. Proc R Soc Lond B 2004;271:251–258.
33 Arakaki N, Miyoshi T, Noda H: *Wolbachia*-mediated parthenogenesis in the predatory thrips *Franklinothrips vespiformis* (Thysanoptera: Insecta). Proc R Soc Lond B Biol Sci 2001;268: 1011–1016.
34 Moritz G: Structure, growth and development; in Lewis T (ed): Thrips As Crop Pests. Cambridge, University Press, 1997, pp 15–63.
35 Weeks AR, Breeuwer JA: *Wolbachia*-induced parthenogenesis in a genus of phytophagous mites. Proc R Soc Lond B Biol Sci 2001;268:2245–2251.
36 Zchori-Fein E, et al: Parthenogenesis-inducing microorganisms in *Aphytis* (Hymenoptera: Aphelinidae). Insect Mol Biol 1995;4:173–178.
37 Pijls JWAM, Van Steenbergen HJ, Van Alphen JJM: Asexuality cured: the relations and differences between sexual and asexual *Apoanagyrus diversicornis*. Heredity 1996;76:402–407.
38 Plantard O, et al: *Wolbachia*-induced thelytoky in the rose gallwasp *Diplolepis spinosissimae* (Giraud) (Hymenoptera: Cynipidae), and its consequences on the genetic structure of its host. Proc R Soc Lond B Biol Sci 1998;265:1075–1080.
39 Pannebakker BA, et al: Cytology of *Wolbachia*-induced parthenogenesis in *Leptopilina clavipes* (Hymenoptera: Figitidae). Genome 2004;47:299–303.
40 Stouthamer R, et al: Molecular identification of microorganisms associated with parthenogenesis. Nature 1993;361:66–68.
41 Arakaki N, Noda H, Yamagishi K: *Wolbachia*-induced parthenogenesis in the egg parasitoid *Telenomus nawai*. Entomol Exp Appl 2000;96:177–184.
42 Stouthamer R, Luck RF, Hamilton WD: Antibiotics cause parthenogenetic *Trichogramma* (Hymenoptera/Trichogrammatidae) to revert to sex. Proc Natl Acad Sci USA 1990;87: 2424–2427.
43 Stouthamer R, et al: Selfish element maintains sex in natural populations of a parasitoid wasp. Proc R Soc Lond B Biol Sci 2001;268:617–622.
44 Werren JH, Zhang W, Guo LR: Evolution and phylogeny of *Wolbachia*: reproductive parasites of arthropods. Proc R Soc Lond B Biol Sci 1995;261:55–63.
45 Stouthamer R, Kazmer DJ: Cytogenetics of microbe associated parthenogenesis, consequences for gene flow in *Trichogramma* wasps. Heredity 1994;73:317–327.
46 Vavre F, de Jong JH, Stouthamer R: Cytogenetic mechanism and genetic consequences of thelytoky in the wasp *Trichogramma cacoeciae*. Heredity 2004;93:592–596.
47 Gottlieb Y, et al: Diploidy restoration in *Wolbachia*-infected *Muscidifurax uniraptor* (Hymenoptera: Pteromalidae). J Invertebr Pathol 2002;81:166–174.
48 Gottlieb Y, Zchori-Fein E: Irreversible thelytokous reproduction in *Muscidifurax uniraptor*. Entomol Exp Appl 2001;100:271–278.
49 Stille B, Dävring L: Meiosis and reproductive strategy in the parthenogenetic gall wasp *Diplolepis rosae* (L.) (Hymenoptera, Cynipidae). Hereditas 1980;92:353–362.
50 Zchori-Fein E, Roush RT, Hunter MS: Male production influenced by antibiotic treatment in *Encarsia formosa* and asexual species. Experientia 1992;48:102–105.
51 Huigens ME, Stouthamer R: Parthenogenesis associated with *Wolbachia*; in Bourtzis K, Miller TA (eds): Insect Symbiosis. Boca Raton, FL, CRC Press, 2003, pp 247–266.
52 Jeong G, Stouthamer R: Genetics of female functional virginity in the parthenogenesis-*Wolbachia* infected parasitoid wasp *Telenomus nawai* (Hymenoptera: Scelionidae). Heredity 2005;94:402–407.
53 Hurst GD: The incidences and evolution of cytoplasmic male killers. Proc R Soc Lond B 1991;244:91–99.
54 Werren JH: The coevolution of autosomal and cytoplasmic sex ratio factors. J Theor Biol 1987;124:317–334.
55 Hurst GD, et al: Male-killing *Wolbachia* in *Drosophila*: a temperature-sensitive trait with a threshold bacterial density. Genetics 2000;156:699–709.
56 Dyer KA, Jaenike J: Evolutionarily stable infection by a male-killing endosymbiont in *Drosophila innubila*: molecular evidence from the host and parasite genomes. Genetics 2004;168:1443–1455.
57 Fialho RF, Stevens L: Male-killing *Wolbachia* in a flour beetle. Proc R Soc Lond B Biol Sci 2000;267:1469–1473.

58 Majerus ME, et al: Multiple causes of male-killing in a single sample of the two-spot ladybird, *Adalia bipunctata* (Coleoptera: coccinellidae) from Moscow. Heredity 2000;84(pt 5):605–609.

59 Jiggins FM, Hurst GD, Majerus ME: Sex-ratio-distorting *Wolbachia* causes sex-role reversal in its butterfly host. Proc R Soc Lond B Biol Sci 2000;267:69–73.

60 Zeh DW, Zeh JA, Bonilla MM: *Wolbachia*, sex ratio bias and apparent male killing in the harlequin beetle riding pseudoscorpion. Heredity 2005;95:41–49.

61 Zeh JA, Zeh DW: Male-killing *Wolbachia* in a live-bearing arthropod: brood abortion as a constraint on the spread of a selfish microbe. J Invertebr Pathol 2006;92:33–38.

62 Kimura MT, et al: Breeding sites of drosophilid flies in and near Sapporo, North Japan, with supplementary notes on the adult feeding habits. Kontyû 1977;45:571–582.

63 Werren JH, Skinner SW, Huger AM: Male-killing bacteria in a parasitic wasp. Science 1986;231:990–992.

64 Breeuwer JA, Jacobs G: *Wolbachia*: intracellular manipulators of mite reproduction. Exp Appl Acarol 1996;20:421–434.

65 Wade MJ, Stevens L: Microorganism mediated reproductive isolation in flour beetles (genus *Tribolium*). Science 1985;227:527–528.

66 Yen JH, Barr AR: The etiological agent of cytoplasmic incompatibility in *Culex pipiens.* J Invertebr Pathol 1973;22:242–250.

67 Moret Y, Juchault P, Rigaud T: *Wolbachia* endosymbiont responsible for cytoplasmic incompatibility in a terrestrial crustacean: effects in natural and foreign hosts. Heredity 2001;86(pt 3):325–332.

68 Brower JH: Cytoplasmic incompatibility: occurence in a stored-product pest *Ephestia cautella.* Ann Entomol Soc Am 1976;69:1011–1015.

69 Reed KM, Werren JH: Induction of paternal genome loss by the paternal-sex-ratio chromosome and cytoplasmic incompatibility bacteria *(Wolbachia)*: a comparative study of early embryonic events. Mol Reprod Dev 1995;40:408–418.

70 Hoshizaki S, Shimada T: PCR-based detection of *Wolbachia*, cytoplasmic incompatibility microorganisms, infected in natural populations of *Laodelphax striatellus* (Homoptera: Delphacidae) in central Japan: has the distribution of *Wolbachia* spread recently? Insect Mol Biol 1995;4:237–243.

71 Kamoda S, et al: *Wolbachia* infection and cytoplasmic incompatibility in the cricket *Teleogryllus taiwanemma.* J Exp Biol 2000;203(pt 16):2503–2509.

72 Caspari E, Watson GS: On the evolutionary importance of cytoplasmic sterility in mosquitoes. Evolution 1959;13:568–570.

73 Fine PE: On the dynamics of symbiote-dependent cytoplasmic incompatibility in culicine mosquitoes. J Invertebr Pathol 1978;31:10–18.

74 Hoffmann AA, Turelli M, Harshman LG: Factors affecting the distribution of cytoplasmic incompatibility in *Drosophila simulans.* Genetics 1990;126:933–948.

75 Turelli M: Evolution of incompatibility-inducing microbes and their hosts. Evolution 1994;48:1500–1513.

76 Turelli M, Hoffmann AA: Cytoplasmic incompatibility in *Drosophila simulans*: dynamics and parameter estimates from natural populations. Genetics 1995;140:1319–1338.

77 Turelli M, Hoffmann AA: Rapid spread of an inherited incompatible factor in *Drosophila.* Nature 1991;353:440–442.

78 Binnington KC, Hoffmann AA: *Wolbachia*-like organisms and cytoplasmic incompatibility in *Drosophila simulans.* J Invertebr Pathol 1989;54:344–352.

79 Snook RR, et al: Offsetting effects of *Wolbachia* infection and heat shock on sperm production in *Drosophila simulans*: analyses of fecundity, fertility and accessory gland proteins. Genetics 2000;155:167–178.

80 Bressac C, Rousset F: The reproductive incompatibility system in *Drosophila simulans*: DAPI-staining analysis of the *Wolbachia* symbionts in sperm cysts. J Invertebr Pathol 1993:61:226–230.

81 Clark ME, et al: The distribution and proliferation of the intracellular bacteria *Wolbachia* during spermatogenesis in *Drosophila.* Mech Dev 2002;111:3–15.

82 Callaini G, et al: Mitotic defects associated with cytoplasmic incompatibility in *Drosophila simulans.* J Invertebr Pathol 1996;67:55–64.

83 Lassy CW, Karr TL: Cytological analysis of fertilization and early embryonic development in incompatible crosses of *Drosophila simulans.* Mech Dev 1996;57:47–58.
84 Clark ME, et al: Widespread prevalence of *Wolbachia* in laboratory stocks and the implications for *Drosophila* research. Genetics 2005;170:1667–1675.
85 Jost E: Genetische Untersuchungen zur kreuzungssterilitat im *Culex pipiens* komplex. Theoret Appl Genet 1970;40:251–256.
86 Tram U, Sullivan W: Role of delayed nuclear envelope breakdown and mitosis in *Wolbachia*-induced cytoplasmic incompatibility. Science 2002;296:1124–1126.
87 O'Neill SO, Karr TL: Bidirectional cytoplasmic incompatibility between conspecific populations of *Drosophila simulans.* Nature 1990;348:178–180.
88 Bordenstein SR, Uy JJ, Werren JH: Host genotype determines cytoplasmic incompatibility type in the haplodiploid genus nasonia. Genetics 2003:164:223–233.
89 Tram U, et al: Paternal chromosome segregation during the first mitotic division determines cytoplasmic incompatibility phenotype. J Cell Sci 2006 (in press).
90 Callaini G, Dallai R, Riparbelli MG: *Wolbachia*-induced delay of paternal chromatin condensation does not prevent maternal chromosomes from entering anaphase in incompatible crosses of *Drosophila simulans.* J Cell Sci 1997;110(pt 2):271–280.
91 Loppin B, Berger F, Couble P: Paternal chromosome incorporation into the zygote nucleus is controlled by maternal haploid in *Drosophila.* Dev Biol 2001;231:383–396.
92 Loppin B, et al: Origin and neofunctionalization of a *Drosophila* paternal effect gene essential for zygote viability. Curr Biol 2005;15:87–93.
93 Loppin B, et al: The maternal effect mutation sésame affects teh formation of the male pronucleus in *Drosophila melanogaster.* Dev Biol 2000;222:392–404.
94 Telschow A, Yamamura N, Werren JH: Bidirectional cytoplasmic incompatibility and the stable coexistence of two *Wolbachia* strains in parapatric host populations. J Theor Biol 2005;235: 265–274.
95 Telschow A, Hammerstein P, Werren JH: The effect of *Wolbachia* versus genetic incompatibilities on reinforcement and speciation. Evolution 2005;59:1607–1619.
96 Telschow A, Hammerstein P, Werren JH: The effect of *Wolbachia* on genetic divergence between populations: models with two way migration. Am Nat 2002;160:S54–S66.
97 Champion de Crespigny FE, Butlin RK, Wedell N: Can cytoplasmic incompatibility inducing *Wolbachia* promote the evolution of mate preferences? J Evol Biol 2005;18:967–977.
98 Giordano R, Jackson JJ, Robertson HM: The role of *Wolbachia* bacteria in reproductive incompatibilities and hybrid zones of *Diabrotica* beetles and *Gryllus* crickets. Proc Natl Acad Sci USA 1997;94:11439–11444.
99 Shoemaker DD, Katju V, Jaenike J: *Wolbachia* and the evolution of reproductive isolation between *Drosophila recens* and *Drosophila subquinaria.* Evolution 1999;53:1157–1164.
100 Bordenstein SR, O'Hara FP, Werren JH: *Wolbachia*-induced incompatibility precedes other hybrid incompatibilities in *Nasonia.* Nature 2001;409:707–710.
101 Coyne JA, Orr HA: Speciation. Sunderland, MA, Sinaur Associates, 2004, pp 276–281.
102 Poinsot D, Charlat S, Mercot H: On the mechanism of *Wolbachia*-induced cytoplasmic incompatibility: confronting the models with the facts. Bioessays 2003;25:259–265.
103 Werren JH: Biology of *Wolbachia.* Annu Rev Entomol 1997;42:587–609.
104 Kose H, Karr TL: Organization of *Wolbachia pipientis* in the *Drosophila* fertilized egg and embryo revealed by an anti-*Wolbachia* monoclonal antibody. Mech Dev 1995;51:275–288.
105 Clark ME, et al: Induced paternal effects mimic cytoplasmic incompatibility in *Drosophila.* Genetics 2006;173:727–734.
106 Stolk C, Stouthamer R: Influence of a cytoplasmic incompatibility-inducing *Wolbachia* on the fitness of the parasitoid wasp *Nasonia vitripennis.* Proc Exp Appl Entomol 1996;7:33–37.
107 Bordenstein SR, Werren JH: Do *Wolbachia* influence fecundity in *Nasonia vitripennis?* Heredity 2000;84(pt 1):54–62.
108 Bourtzis K, et al: *Wolbachia* infection and cytoplasmic incompatibility in *Drosophila* species. Genetics 1996;144:1063–1073.
109 Poinsot D, et al: *Wolbachia* transfer from *Drosophila melanogaster* into *D. simulans*: host effect and cytoplasmic incompatibility relationships. Genetics 1998;150:227–237.

110 Bourtzis K, et al: A prokaryotic dnaA sequence in *Drosophila melanogaster: Wolbachia* infection and cytoplasmic incompatibility among laboratory strains. Insect Mol Biol 1994;3:131–142.
111 Hoffmann AA, Clancy DJ, Merton E: Cytoplasmic incompatibility in Australian populations of *Drosophila melanogaster.* Genetics 1994;136:993–999.
112 Reynolds KT, Thomson LJ, Hoffmann AA: The effects of host age, host nuclear background and temperature on phenotypic effects of the virulent *Wolbachia* strain popcorn in *Drosophila melanogaster.* Genetics 2003;164:1027–1034.
113 Starr DJ, Cline TW: A host parasite interaction rescues *Drosophila* oogenesis defects. Nature 2002;418:76–79.
114 Rigaud T, Juchault P, Mocquard JP: The evolution of sex determination in isopod crustaceans. Bioessays 1997;19:409–416.
115 Stouthamer R, Breeuwer JA, Hurst GD: *Wolbachia pipientis*: microbial manipulator of arthropod reproduction. Annu Rev Microbiol 1999;53:71–102.
116 Traut W, et al: Phylogeny of sex-determining gene Sex-Lethal in insects. Genome 2006;49:254–262.
117 Meise M, et al: Sex-lethal, the master sex-determining gene in *Drosophila,* is not sex-specifically regulated in *Musca domestica.* Development 1998;125:1487–1494.
118 Dedeine F, et al: Removing symbiotic *Wolbachia* bacteria specifically inhibits oogenesis in a parasitic wasp. Proc Natl Acad Sci USA 2001;98:6247–6252.
119 Dedeine F, et al: Intra-individual coexistence of a *Wolbachia* strain required for host oogenesis with two strains inducing cytoplasmic incompatibility in the wasp *Asobara tabida.* Evolution 2004;58:2167–2174.
120 Hoerauf A, et al: Tetracycline therapy targets intracellular bacteria in the filarial nematode *Litomosoides sigmodontis* and results in filarial infertility. J Clin Invest 1999;103:11–18.
121 Zchori-Fein E, Chandresh B, Harari AR: Oogenesis in the date stone beetle, *Coccotrypes dactyliperda,* depends on symbiotic bacteria. Physiol Entomol 2006;31:164–169.
122 Vela F, et al: Genetic conflicts over sex ratio: mite-endosymbiont interactions. Am Nat 2003;36: 254–266.
123 Hoffmann AA, Hercus M, Dagher H: Population dynamics of the *Wolbachia* infection causing cytoplasmic incompatibility in *Drosophila melanogaster.* Genetics 1998;148:221–231.
124 Riegler M, et al: Evidence for a global *Wolbachia* replacement in *Drosophila melanogaster.* Curr Biol 2005;15:1428–1433.
125 Fry AJ, Rand DM: *Wolbachia* interactions that determine *Drosophila melanogaster* survival. Evolution Int J Org Evolution 2002;56:1976–1981.
126 Fry AJ, Palmer MR, Rand DM: Variable fitness effects of *Wolbachia* infection in *Drosophila melanogaster.* Heredity 2004;93:379–389.
127 Driver C, Tawadros N: Cytoplasmic genomes that confer additional longevity in *Drosophila melanogaster.* Biogerontology 2000;1:255–260.
128 Driver C, Georgiou A, Georgiou G: The contribution by mitochondrially induced oxidative damage to aging in *Drosophila melanogaster.* Biogerontology 2004;5:185–192.
129 Min KT, Benzer S: *Wolbachia,* normally a symbiont of *Drosophila,* can be virulent, causing degeneration and early death. Proc Natl Acad Sci USA 1997;94:10792–10796.
130 Harcombe W, Hoffmann AA: *Wolbachia* effects in *Drosophila melanogaster*: in search of fitness benefits. J Invertebr Pathol 2004;87:45–50.
131 Dobson SL: Evolution of *Wolbachia* cytoplasmic incompatibility types. Evolution 2004;58: 2156–2166.
132 Bordenstein SR, Werren JH: Effects of A and B *Wolbachia* and host genotype on interspecies cytoplasmic incompatibility in *Nasonia.* Genetics 1998;148:1833–1844.
133 Koukou K, et al: Influence of antibiotic treatment and *Wolbachia* curing on sexual isolation among *Drosophila melanogaster* cage populations. Evolution 2006;60:87–96.
134 Bourtzis K, Pettigrew MM, O'Neill SL: *Wolbachia* neither induces nor suppresses transcripts encoding antimicrobial peptides. Insect Mol Biol 2000;9:635–639.
135 Fytrou A, et al: *Wolbachia* infection suppresses both host defence and parasitoid counter-defence. Proc Biol Sci 2006;273:791–796.
136 Wade MJ, Chang NW: *In*creased male fertility in *Tribolium confusum* beetles after infection with the intracellular parasite *Wolbachia.* Nature 1995;373:72–74.

137 Champion de Crespigny FE, Wedell N: *Wolbachia* infection reduces sperm competitive ability in an insect. Proc R Soc Lond B Biol Sci 2006;273:1455–1458.

138 Dunn AK, et al: Evolutionary ecology of vertically transmitted parasites: strategies of transovarial transmission of a microsporidian sex ratio distorter in *Gammarus duebeni.* Parasitology 1996;111: S91–S109.

139 Terry RS, Dunn AM, Smith JE: Cellular distribution of a feminizing microsporidian parasite: strategy for transovarial transmission. Parasitology 1997;115:157–163.

140 Rodgers-Gray TP, et al: Mechanisms of parasite-induced sex reversal in *Gammarus duebeni.* Int J Parasitol 2004;34:747–753.

141 Weeks AR, Marec F, Breeuwer JA: A mite species that consists entirely of haploid females. Science 2001;292:2479–2482.

142 Hagimori T, et al: The first finding of a *Rickettsia* bacterium associated with parthenogenesis induction among insects. Curr Microbiol 2006;52: 97–101.

143 Zchori-Fein E, et al: A newly discovered bacterium associated with parthenogenesis and a change in host selection behavior in parasitoid wasps. Proc Natl Acad Sci USA 2001;98:12555–12560.

144 Zchori-Fein E, et al: Characterization of a 'Bacteroidetes' symbiont in *Encarsia* wasps (Hymenoptera: Aphelinidae): proposal of '*Candidatus Cardinium hertigii*'. Int J Syst Evol Microbiol 2004;54:961–968.

145 Williamson DL, et al: *Spiroplasma poulsonii* sp. Nov., a new species associated with male-lethality in *Drosophila willistoni*, a neotropical species of fruit fly. Int J Syst Bacteriol 1999;49: 611–618.

146 Montenegro H, et al: Male-killing *Spiroplasma* naturally infecting *Drosophila melanogaster.* Insect Mol Biol 2005;14:281–287.

147 Tinsley MC, Majerus ME: A new male-killing parasitism: *Spiroplasma* bacteria infect the ladybird beetle *Anisosticta novemdecimpunctata* (Coleoptera: Coccinellidae). Parasitology 2006;132: 757–765.

148 Hurst GD, et al: Invasion of one insect species, *Adalia bipunctata*, by two different male-killing bacteria. Insect Mol Biol 1999;8:133–139.

149 Majerus TM, et al: Molecular identification of a male-killing agent in the ladybird *Harmonia axyridis* (Pallas) (Coleoptera: Coccinellidae). Insect Mol Biol 1999;8:551–555.

150 Jiggins FM, et al: The Butterfly *Danaus chrysippus* is infected by the male-killing spiroplasma bacterium. Parasitology 2000;120:439–446.

151 Werren JH, et al: Rickettsial relative associate with male-killing in the ladybird beetle *(Adalia bipunctata).* J Bacteriol 1994;176:388–394.

152 von der Schulenburg JH, et al: Incidence of male-killing *Rickettsia* spp. (alpha-proteobacteria) in the ten-spot lady beetle *Adalia decempunctata* L. (Coleoptera: Coccinellidae). Appl Environ Microbiol 2001;67:270–277.

153 Lawson ET, et al: *Rickettsia* associated with male-killing in a buprestid beetle. Heredity 2001;86: 497–505.

154 Majerus MEN: Sex Wars: Genes, Bacteria, and Biased Sex Ratios. Princeton, New Jersey, Princeton University Press, 2003, pp 137–143.

155 Hurst GD, et al: *Adonia variegata* (Coleoptera: Coccinellidae) bears maternally inherited flavobacteria that kill males only. Parasitology 1999;118:125–134.

156 Hurst GD, et al: Male-killing bacterium in a fifth ladybird beetle, *Coleomigilla maculata* (Coleoptera: Coccinellidae). Heredity 1996;77:177–185.

157 Dyson EA, Kamath MK, Hurst GD: *Wolbachia* infection associated with all-female broods in *Hypolimnas bolina* (Lepidoptera: Nymphalidae): evidence for horizontal transmission of a butterfly male killer. Heredity 2002;88:166–171.

158 Hunter MS, Perlman SJ, Kelly SE: A bacterial symbiont in the Bacteriodetes induces cytoplasmic incompatibility in the parasitoid wasp *Encarsia pergandiella.* Proc Biol Sci 2003;270:2185–2190.

159 Weeks AR, Velten R, Stouthamer R: Incidence of a new sex-ratio-distorting endosymbiotic bacterium among arthropods. Proc R Soc Lond B Biol Sci 2003;270:1857–1865.

160 Peacock WJ, Erickson J: An indicator of polarity in the spermatocyte? Drosoph Inf Serv 1964;39: 107–108.

161 Veneti Z, et al: Heads or tails: host-parasite interactions in the *Drosophila-Wolbachia* system. Appl Environ Microbiol 2004;70:5366–5372.
162 Dudkina NV, Kiseleva EV: Structural organization and distribution of the symbiotic bacteria *Wolbachia* during spermatogenesis of *Drosophila simulans.* Ontogenez 2005;36:41–50.
163 Fuller MT: Spermatogenesis; in Bates M, Martinez Arias A (eds): The Development of *Drosophila melanogaster*. Cold Spring Harbor, NY, Cold Spring Harbor Press, 1993, pp 71–147.
164 Clark ME, et al: *Wolbachia* distribution and cytoplasmic incompatibility during sperm development: the cyst as the basic cellular unit of CI expression. Mech Dev 2003;120:185–198.
165 Dudkina NV, Voronin DA, Kiseleva EV: Structural organization and distribution of symbiotic bacteria *Wolbachia* in early embryos and ovaries of *Drosophila melanogaster* and *D. simulans.* Tsitologiia 2004;46:208–220.
166 Wolstenholme DR: A DNA and RNA-containing cytoplasmic body in *Drosophila melanogaster* and its relation to flies. Genetics 1965;52:949–975.
167 Ferree PM, et al: *Wolbachia* utilizes host microtubules and dynein for anterior localization in the *Drosophila* oocyte. PLoS Pathog 2005;1:e14.
168 Voronin DA, Dudkina NV, Kiseleva EV: A new form of symbiotic bacteria *Wolbachia* found in the endoplasmic reticulum of early embryos of *Drosophila melanogaster.* Dokl Biol Sci 2004;396: 227–229.
169 Callaini G, Riparbelli MG, Dallai R: The distribution of cytoplasmic bacteria in early *Drosophila* embryo is mediated by astral microtubules. J Cell Sci 1994;107:673–682.
170 Breeuwer JA, Werren JH: Microorganisms associated with chromosome destruction and reproductive isolation between two insect species. Nature 1990;346:558–560.
171 Zchori-Fein E, Roush RT, Rosen D: Distribution of parthenogenesis-inducing symbionts in ovaries and eggs of *Aphytis* (Hymentoptera: Aphelinidae). Curr Microbiol 1998;36:1–8.

Michael E. Clark
Department of Biology, University of Rochester
Rochester, NY 14627 (USA)
Tel. +1 585 275 3889, Fax +1 585 275 2070, E-Mail mclark11@mail.rochester.edu

Hoerauf A, Rao RU (eds): Wolbachia.
Issues Infect Dis. Basel, Karger, 2007, vol 5, pp 124–132

Wolbachia and Its Importance in Veterinary Filariasis

Laura Kramer[a], *John W. McCall*[c], *Giulio Grandi*[a], *Claudio Genchi*[b]

[a]Department of Animal Production, College of Veterinary Medicine, University of Parma, Parma, and [b]Department of Veterinary Pathology and Parasitology, College of Veterinary Medicine, University of Milan, Milan, Italy; [c]Department of Infectious Diseases, College of Veterinary Medicine, University of Georgia, Athens, Ga., USA

Abstract

Filarial worms are generally of little veterinary importance, with the exception of *Dirofilaria immitis*, the agent of canine and feline heartworm disease. Another filarial nematode, *Onchocerca ochengi*, causes subcutaneous filariasis in cattle and has become an increasingly important animal model for the study of human oncocerciasis and in particular for the evaluation of therapeutic strategies that target *Wolbachia*. This chapter reviews previous and current work carried out to define the importance of *Wolbachia* in the pathogenesis, diagnosis and treatment of these infections and how the results of such work may also contribute to the study of *Wolbachia* in human filariasis.

When Sironi et al. [1] identified the intracellular bacteria within *Dirofilaria immitis* as belonging to the genus *Wolbachia*, a new chapter opened in the history of human and animal filarial infection. The significance of this study was so great that many research groups immediately recognized the potential of *Wolbachia* as a therapeutic target for human filarial disease, as illustrated elsewhere in this volume.

Several groups, however, have continued studying the role of *Wolbachia* in those filarial nematodes that cause infection/disease in domestic animals. Many of these studies have been aimed at using animal infection as a model for evaluating the effects of antibiotic treatment on worm fecundity and survival, but others have attempted to look more closely at the role of *Wolbachia* in the pathogenesis, immune response and diagnosis of the diseases they cause in their definitive hosts. The search for effective therapeutic protocols against infection is of primary importance also in veterinary medicine and the current status of knowledge on *Wolbachia* in veterinary filariasis will be the focus of this chapter.

D. immitis and Heartworm Disease

Heartworm disease is a parasitic infection caused by *D. immitis* in canine and feline populations worldwide. The adult worms live in the pulmonary arteries where mature females release first-stage larvae (microfilariae) into the bloodstream. These are taken up by an arthropod vector (several species of mosquitoes are competent vectors of *D. immitis*) where they develop to the third stage, which becomes infective (L_3). These L_3 are then deposited in the dermis of the final host when the infected mosquito bites the host and enters the body via the puncture wound made by the mosquito's skin-piercing mouthparts. After several months of migration and maturation, they reach the pulmonary arteries where they continue to grow, mate and produce microfilariae. Heartworm infection in dogs usually presents as a chronic disease. Initially, the pulmonary vasculature is affected and later on the lung itself, and finally the right chambers of the heart. It seems that this final migration to the right chamber of the heart may have several causes: either the embolization of the pulmonary arteries, or the release of substances from dead worms. Both of these causes have been investigated: while the embolization of the lung vessels causes permanent migration of the worms to the right atrium, the inoculation of dead worm extract elicits temporary migration of the worms [2].

The initial inflammatory reaction that occurs in the walls of the pulmonary vasculature is critical in the development of the entire disease process [3]. Feline infection is diagnosed with increasing frequency in areas where the disease is endemic in canines [4]. However, the development of the parasite and the clinical findings in cats are different from those which occur in dogs. The development of the parasite takes longer compared to dogs and most infections are amicrofilaremic. Additionally, the parasite burden is low and the infection is generally asymptomatic, although some cats present with severe disease or even sudden death in the presence of a small number of adult worms (one to three) [4–6].

Wolbachia in *D. immitis*

Initial descriptions of bacterial-like structures using electron microscopy [7] and more recent studies by immunohistology have provided a comprehensive description of the distribution of *Wolbachia* in *D. immitis* [8, 9]. In adult

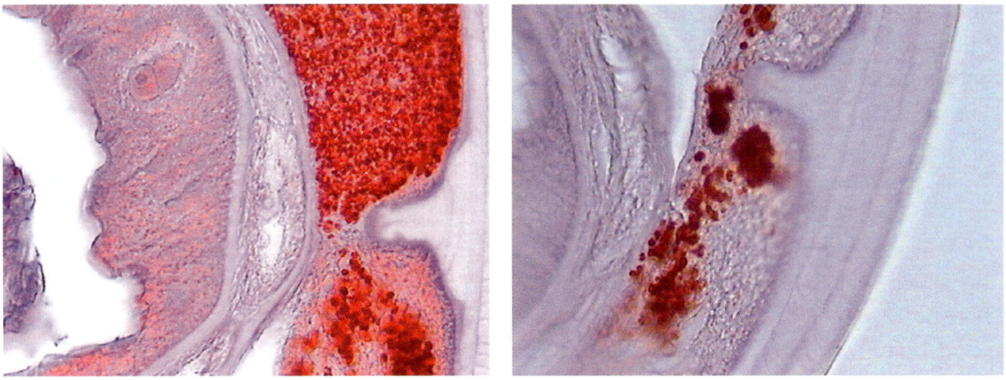

a b

Fig. 1. Hypodermal chord of a female *D. immitis*. Wolbachia distribution is not always uniform and it is possible, in a cross-section like the one shown, to find the entire hypodermal cell filled (***a***) or to observe only a few scattered bacteria (***b***). Immunohistochemical staining with an anti-WSP polyclonal serum and ABC-HSP method, as described in Kramer et al. [9]. Magnification 100×.

D. immitis, *Wolbachia* is predominantly found throughout the hypodermal cells of the lateral cords (fig. 1). The bacteria occur within host-derived vacuoles in variously sized discrete groups ranging from a few organisms, often clustered around hypodermal nuclei, to areas where they almost completely fill the cellular environment reminiscent of bacteriocyte-like structures. In female heartworms, *Wolbachia* is also present in the ovaries, oocytes and developing embryonic stages within the uteri (fig. 2), whereas they have not been demonstrated in the male reproductive system [7]. The most recent phylogenic analysis places the *Wolbachia* of *D. immitis* in clade C, together with the *Wolbachia* from *Dirofilaria repens* (the cause of subcutaneous dirofilariasis) and the *Wolbachia* from *Onchocerca* spp. [10]. To date, two proteins from the *Wolbachia* of *D. immitis* have been produced in recombinant form: *Wolbachia* surface protein (WSP) [11] and GroEL [12].

Wolbachia in Heartworm-Infected Animals

Most evidence indicates that the filarial-infected host comes into contact with *Wolbachia* following the death of worms (macro-microfilariae through natural attrition, microfilarial turnover and/or pharmacological intervention). However, Kozek [13] has recently hypothesized that living worms may release *Wolbachia* and/or their products, possibly from uterine debris, which promote

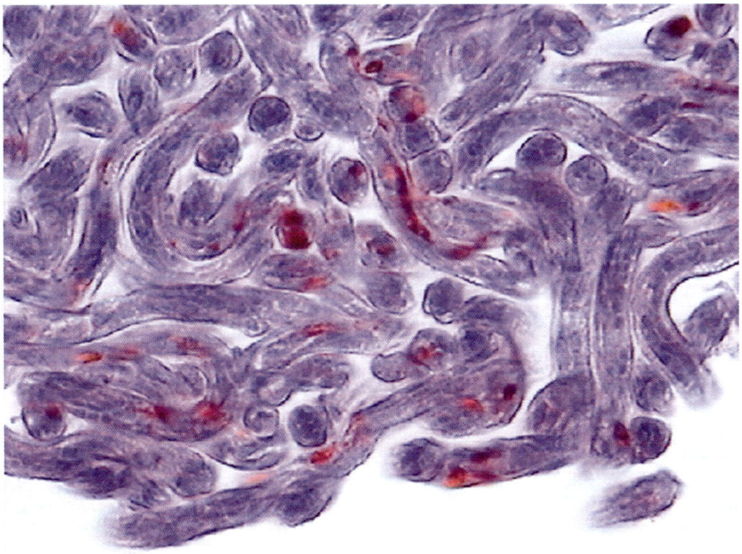

Fig. 2. Uterus of a female *D. immitis*. Notice how the stretched microfilariae contain numerous *Wolbachia*. Immunohistochemical staining with an anti-WSP polyclonal serum and ABC-HSP method, as described in Kramer et al. [9]. Magnification 100×.

inflammatory responses adjacent to the worms. Interestingly, a major surface protein of *Wolbachia* from *D. immitis* has been shown to provoke chemiokinesis and IL-8 production in canine neutrophils in vitro [14].

We recently tested the hypothesis that *D. immitis*-infected dogs also come into contact with *Wolbachia* either through microfilarial turnover or natural death of adult worms [15]. Immunoglobulin G (total IgG, IgG1, IgG2) production against and immunohistochemical staining of tissues for the WSP from dogs with natural heartworm infection were evaluated. All infected dogs had significantly higher total anti-WSP IgG levels compared to healthy controls. Interestingly, WSP was recognized by the IgG2 subclass in both microfilariemic dogs and in dogs with no circulating microfilariae (occult infection). However, microfilariemic dogs also produced IgG1 antibodies. Positive staining for WSP was observed in lungs, liver and kidneys, in particular in glomerular capillaries of naturally infected dogs who had died from heartworm disease. Interestingly, immune-complex glomerulonephritis is a frequent complication of heartworm disease and the localization of WSP in glomeruli is suggestive of a role for *Wolbachia* in renal pathology [16]. It has been reported that infection in dogs with *Ehrlichia canis*, a bacteria closely related to *Wolbachia*, features immune-complex formation that may be responsible for renal lesions. These results showed for the first time that *Wolbachia* is recognized specifically by *D. immitis*-infected dogs and that the bacteria is released into host tissue. Furthermore, microfilariemia status appeared to affect immune responses to this endosymbiont.

Bazzocchi et al. [17] reported the specific recognition of WSP in Western blotting with serum of experimentally infected cats. In a more recent study, the antibody response against specific molecules of *D. immitis* and *Wolbachia* endosymbionts in both naturally and experimentally infected cats with and without larvicidal (ivermectin) treatment, was evaluated [18]. Increased antibody production against filarial antigens and WSP was observed in experimentally infected cats without treatment. However, in experimentally infected cats treated with a larvicidal drug, there was a transient increase in anti-*D. immitis* IgG that decreased dramatically in association with the death of the larvae, while the anti-WSP IgG response increased constantly until the end of the experiment (6 months). The immune response to *Wolbachia* antigens was detected as early as 2 months after infection, before detection of specific antibodies against *D. immitis* antigens. These findings suggest that *Wolbachia* also plays an important role in the immune response to heartworm infection in cats and may also have diagnostic value. An intriguing question arises, however: is it possible that this intense immune response to *Wolbachia* is characteristic of resistance to infection in this host?

Perhaps one of the most interesting results seen so far with infection by *D. immitis* concerns human dirofilariasis. Simòn et al. [19] have reported specific humoral recognition of WSP in patients with pulmonary nodules due to migration of *D. immitis* and have suggested the use of this antibody response in the differential diagnosis of the disease. Indeed, people living in areas endemic for *D. immitis* are at risk of infection. Humans are, however, 'dead-end hosts', since larvae do not normally develop into the adult stage in humans. Human pulmonary dirofilariasis develops when the nematode travels to the lung, lodges in a small branch of the pulmonary artery, dies and embolizes. In endemic areas, clinically healthy people are frequently found positive for antibodies against *D. immitis* antigens [20] and the high percentage (26–37%) of seroprevalence in healthy people in areas of canine heartworm endemicity hampers the serological diagnosis of human pulmonary dirofilariasis. Results of the study showed that the IgG response against the WSP of *D. immitis* is consistently detectable only in patients with pulmonary nodules due to the parasite, suggesting that the surface protein of *Wolbachia* endosymbionts stimulates the host immune system only after the death of immature adult worms in the small branches of pulmonary arteries, or at least when the development of *D. immitis* has progressed to a stage at which nematode death can lead to the release of a sufficient amount of bacteria.

The Effect of Antibiotics on D. immitis

Wolbachia population dynamics may explain the differential activity of bacteriostatic antibiotic treatment on distinct developmental stages of filarial worms, in which larval and embryonic development are associated with rapidly dividing bacteria and are affected soon after antibiotic treatment, whereas the more slowly dividing populations in adults take longer to deplete the bacterial population and for the consequences to show. Bandi et al. [8] reported that *D. immitis* adults taken from naturally infected dogs that had been treated with $20\,mg \cdot kg^{-1} \cdot day^{-1}$ of doxycycline for 30 days were lively and motile, exactly like their control counterparts. Furthermore, there was no difference in microfilarial concentration between treated and control dogs. However, when uterine content of these worms was examined, there was a dramatic decrease in the number of pretzels and stretched microfilariae, indicating that bacteriostatic antibiotic treatment was able to block embryogenesis.

There are little data concerning the effects of antibiotic treatment of dogs with natural heartworm disease. We know, however, that such treatment drastically reduces *Wolbachia* loads in *D. immitis* [8]. We are currently carrying out a clinical trial in experimentally infected dogs: preliminary evidence suggests that doxycycline is able to inhibit embryogenesis and gradually deplete the population of circulating microfilariae [McCall, pers. commun.]. Use of antibiotics during heartworm disease merits further study to determine if treatment (1) is able to reduce inflammatory response to infection by reducing *Wolbachia* load, (2) may help to alleviate the side effects of adulticide therapy by reducing *Wolbachia* load and (3) is adulticide. This last point is particularly important. The current American Heartworm Society Guidelines [21] recommend the use of melarsomine hydrochloride for adulticide therapy in dogs. The American Heartworm Society recommends that this drug be given as a single intramuscular injection, followed at least 1 month later by two intramuscular injections 24 h apart. It can cause potentially life-threatening side effects, the most severe being due to pulmonary thromboembolism [21]. An alternative macrofilaricide protocol may indeed prove beneficial in many cases of canine heartworm disease. Furthermore, there are currently no drugs registered for adulticide therapy in cats.

Onchocerca ochengi

O. ochengi is a filarial nematode that has been widely studied as an important model for human infection by *Onchocerca volvulus*, the agent of human onchocerciasis (river blindness). *O. ochengi* infection causes subcutaneous nodules in cattle. There are several features of *O. ochengi* infection that make it an excellent model for human disease [22]: the two species are closely related, as shown by morphological, biological and phylogenic analyses. They share the same biological vector and, perhaps more intriguing, they stimulate cross-reactive immune responses. For example, Wahl et al. [23] reported that a high prevalence of *O. ochengi* infection in the vector *Simulium damnosum* may have protected humans in an area endemic for *O. volvulus* due to a 'natural heterologous vaccination' by the large number of *O. ochengi*-L3 transmitted to people annually by the anthropo-boophilic vector.

Perhaps more important, however, is the use of *O. ochengi*-infected cattle for evaluating the efficacy of different therapeutic regimes which may then be applicable to *O. volvulus* infection in man. And here is where the most recent advances have been made concerning *Wolbachia*. In fact, the first report of the potential macrofilaricide effects of doxycycline were from studies conducted in cattle infected with *O. ochengi*. It was the casual observation that in cattle with numerous skin nodules due to *O. ochengi*, oxytetracycline treatment for an unrelated infection led to regression of the nodules, that stimulated interest in this model. Langworthy et al. [24] showed that intermittent therapy with oxytetracycline for a 6-month period caused the death of adult worms by 9 months after treatment, and they also showed that worm death was preceded by elimination of *Wolbachia*. A more recent study [25] has reported that different treatment regimes are more or less successful in eliminating the bacteria from the skin-dwelling worms and that this may affect the ability of such treatment to kill adult worms. The implications of these studies for eventual adulticide therapy in human onchocerciasis are indeed overwhelming.

Conclusions

Even though filarial worms have limited importance in veterinary medicine, we have seen how the discovery of *Wolbachia*, this 'bug within the bug' has also contributed to innovative research and advances in one of the most important of these, *D. immitis*. Much remains to be done, in particular future studies on the therapeutic benefits of antibiotic treatment aimed at eliminating *Wolbachia* load within adult worms.

References

1 Sironi M, Bandi C, Sacchi L, Di Sacco B, Damiani G, Genchi C: Molecular evidence for a close relative of the arthropod endosymbiont *Wolbachia* in a filarial worm. Mol Biochem Parasitol 1995;74:223–227.

2 Kitagawa H, Sasaki Y, Ishihara K, Kawakami M: Heartworm migration toward right atrium following artificial pulmonary arterial embolism or injection of heartworm body fluid. Jpn J Vet Sci 1990;52:591–599.

3 Furlanello T, Caldin A, Vezzoni A, Venco L, Kitagawa H: La filariosi cardiopolmonare del cane e del gatto. Cremona, SCIVAC, 1998.
4 Atkins CE, De Francesco TC, Coats JR, Sidley JA, Keene BW: Heartworm infection in cats: 50 cases. J Am Vet Med Assoc 2000;217:355–358.
5 Genchi C, Guerrero J, Di Sacco B, Formaggini L: Prevalence of *Dirofilaria immitis* infection in Italian cats; in Soll MD (ed): Proceedings of the Heartworm Symposium, Batavia, 1992, pp 97–102.
6 McCall JW, Calvert CA, Rawlings CA: Heartworm infection in cats: a life-threatening disease. Vet Med 1994;89:639–647.
7 Sacchi L, Corona S, Casiraghi M, Bandi C: Does fertilization in the filarial nematode *Dirofilaria immitis* occur through endocytosis of spermatozoa? Parasitology 2002;124:87–95.
8 Bandi C, McCall JW, Genchi C, Corona S, Venco L, Sacchi L: Effects of tetracycline on the filarial worms *Brugia pahangi* and *Dirofilaria immitis* and their bacterial endosymbionts *Wolbachia*. Int J Parasitol 1999;29:357–364.
9 Kramer LH, Passeri B, Corona S, Simoncini L, Casiraghi M: Immunohistochemical/immunogold detection and distribution of the endosymbiont *Wolbachia* of *Dirofilaria immitis* and *Brugia pahangi* using a polyclonal antiserum raised against WSP (*Wolbachia* Surface Protein). Parasitol Res 2003;89:381–386.
10 Casiraghi M, Bordenstein SR, Baldo L, Lo N, Beninati T, Wernegreen JJ, Werren JH, Bandi C: Phylogeny of *Wolbachia pipientis* based on gltA, groEL and ftsZ gene sequences: clustering of arthropod and nematode symbionts in the F supergroup, and evidence for further diversity in the *Wolbachia* tree. Microbiology 2005;151:4015–4022.
11 Bazzocchi C, Jamnongluk W, O'Neill S, Anderson TJC, Genchi C, Bandi C: wsp gene sequences from the *Wolbachia* of filarial nematodes. Curr Microbiol 2000;41:96–100
12 Bazzocchi C, Lecchi C, Kramer LH, Genchi C, Bandi C: Sequencing of the complete gene coding for the GroEL of the *Wolbachia* of *Dirofilaria immitis* and expression and purification of the recombinant protein. Parassitologia 2004;46:307–310.
13 Kozek WJ: What is new in the *Wolbachia/Dirofilaria* interaction? Vet Parasitol 2005;133: 127–132.
14 Bazzocchi C, Genchi C, Paltrinieri S, Lecchi C, Mortarino M, Bandi C: Immunological role of the endosymbionts of *Dirofilaria immitis*: the *Wolbachia* surface protein activates canine neutrophils with production of IL-8. Vet Parasitol 2003;117:73–83.
15 Kramer LH, Tamarozzi F, Morchon R, Lopez-Belmonte J, Marcos- Atxutegi C, Martin-Pacho R, Simon F: Immune response to and tissue localization of the *Wolbachia* surface protein (WSP) in dogs with natural heartworm (*Dirofilaria immitis*) infection. Vet Immunol Immunopathol 2005;106: 303–308.
16 Nakagaki K, Nogami S, Hayashi Y, Hammerberg B, Tanaka H, Ohishi I: *Dirofilaria immitis*: detection of parasite-specific antigen by monoclonal antibodies in glomerulonephritis in infected dogs. Parasitol Res 1993;79:49–54.
17 Bazzocchi C, Ceciliani F, McCall JW, Ricci I, Genchi C, Bandi C: Antigenic role of the endosymbionts of filarial nematodes: IgG response against the *Wolbachia* surface protein in cats infected with *Dirofilaria immitis*. Proc R Soc Lond B 2000;267:1–6.
18 Morchon R, Ferreira AC, Martin-Pacho J, Montoya A, Mortarino M, Genchi C, Simon F: Specific IgG antibody response against antigens of *Dirofilaria immitis* and its *Wolbachia* endosymbiont bacterium in cats with natural and experimental infections. Vet Parasitol 2004;125:313–321.
19 Simón F, Prieto G, Morchón R, Bazzocchi C, Bandi C, Genchi C: IgG antibodies against the endosymbionts of filarial nematodos (*Wolbachia*) in patients with pulmonary dirofilariosis. Clín Diag Lab Immunol 2003;10:180–181.
20 Prieto G, Cancrini G, Muro A, Genchi C, Simon F: Seroepidemiology of *Dirofilaria immitis* and *Dirofilaria repens* in humans from three areas of southern Europe. Res Rev Parasitol 2000;60: 95–98.
21 Nelson CT, McCall JW, Rubin SB, Buzhardt LF, Dorion DW, Graham W, Longhofer SL, Guerrero J, Robertson-Plouch C, Paul A; Executive Board of the American Heartworm Society: Guidelines for the diagnosis, prevention and management of heartworm (*Dirofilaria immitis*) infection in dogs. Vet Parasitol 2005;133:255–266.

22 Trees AJ, Graham SP, Renz A, Bianco AE, Tanya V: *Onchocerca ochengi* infections in cattle asa model for human onchocerciasis: recent developments. Parasitology 2000;120:S133–S142.
23 Wahl G, Enyong P, Ngosso A, Schibel JM, Moyou R, Tubbesing H, Ekale D, Renz A: *Onchocerca ochengi*: epidemiological evidence of cross-protection against *Onchocerca volvulus* in man. Parasitology 1998;116:349–362.
24 Langworthy NG, Renz A, Mackenstedt U, Henkle-Duhrsen K, de Bronsvoort MB, Tanya VN, Donnelly MJ, Trees AJ: Macrofilaricidal activity of tetracycline against the filarial nematode *Onchocerca ochengi*: elimination of *Wolbachia* precedes worm death and suggests a dependent relationship. Proc R Soc Lond B 2000;267:1063–1069.
25 Gilbert J, Nfon CK, Makepeace BL, Njongmeta LM, Hastings IM, Pfarr KM, Renz A, Tanya VN, Trees AJ: Antibiotic chemotherapy of onchocerciasis: in a bovine model, killing of adult parasites requires a sustained depletion of endosymbiotic bacteria (*Wolbachia* species). J Infect Dis 2005;192:1483–1493.

Laura H. Kramer
Dipt. di Produzioni Animali, Università di Parma
Via del Taglio 8
IT–43100 Parma (Italy)
Tel. +39 0521 902 715, Fax +39 0521 902 770, E-Mail kramerlh@unipr.it

Hoerauf A, Rao RU (eds): Wolbachia.
Issues Infect Dis. Basel, Karger, 2007, vol 5, pp 133–145

Wolbachia and *Onchocerca volvulus*: Pathogenesis of River Blindness

Katrin Daehnel[a], *Amy G. Hise*[b], *Illona Gillette-Ferguson*[a], *Eric Pearlman*[a,b]

[a]Department of Ophthalmology and [b]The Center for Global Health and Diseases, Case Western Reserve University, Cleveland, Ohio, USA

Abstract

Over 150 million individuals worldwide are infected with filarial nematodes, which include *Wuchereria bancrofti* and *Brugia malayi* that cause lymphatic filariasis, and *Onchocerca volvulus*, which causes onchocerciasis (river blindness). These nematodes all harbor endosymbiotic *Wolbachia* bacteria throughout their life cycles, and it has become increasingly clear that *Wolbachia* have an important role in the pathogenesis of disease in the human host. This review discusses the evidence supporting the role of *Wolbachia* in the pathogenesis of river blindness, and the critical role of the Toll-like receptor pathway in the host immune response to these bacteria.

Filarial nematodes are the only group of nematodes that harbor the endosymbiont *Wolbachia* species, and are also the only nematodes that have an obligate insect vector [1–3]. Most species of filarial nematodes are infected with *Wolbachia pipientis*, including those mentioned above that cause lymphatic filariasis and onchocerciasis [1–3]. Of some 27 species of filariae, there are few that do not harbor *Wolbachia*, including *Loa loa*, which causes human loiasis, and the deer parasite *Acanthocheilonema viteae* used in experiments described below [1–4], and phylogenetic evidence indicates that *Wolbachia* was lost from these species in the process of evolution rather than never harboring *Wolbachia* [5]. Indeed, phylogenetic analysis indicates that transfer of arthropod *Wolbachia* to nematodes occurred more than once over the course of evolution [6]. *Wolbachia* reside in host-derived vacuoles within hypodermal cord cells and within ovarian tissues and developing embryos in the uterus [1].

Bacteria are transmitted transovarially, and antibiotic treatment of infected individuals inhibits worm development, blocks embryogenesis, and kills the embryos indicating that *Wolbachia* have an essential role in nematode development [1, 7, 8]. Antibiotic treatment results in reduction of circulating larvae, and thereby reduces transmission of the parasites [9–11].

The life cycle of all filarial nematodes includes transmission through insect vectors, including mosquitoes for *Wuchereria bancrofti* and *Brugia malayi*, and blackflies for *Onchocerca volvulus*. First-stage larvae (microfilariae or L1) ingested during a blood meal migrate through the insect gut, the thorax and into the salivary gland undergoing two molts to the third-stage larvae (L3) that then enter the mammalian host during a second blood meal. In the mammalian host, the larvae develop to L4 stage and then adult males and females. In lymphatic filariasis, adult worms reside within the large lymphatic vessels, whereas in onchocerciasis, adult male and female worms collect in subcutaneous nodules. Using quantitative PCR for *Wolbachia* surface protein (WSP) gene, which is present as a single copy per organism, the number of bacteria per worm was found to be similar in all insect stages of *B. malayi* [12]. However, within 7 days in the mammalian host, bacterial numbers increased 600-fold and showed a high ratio of *Wolbachia*/nematode DNA in L4 larvae, indicating rapid bacterial replication within the worms. This number of *Wolbachia* is maintained in adult males, but increases in females during embryogenesis [12].

Several recent reviews have described in detail the interactions between *Wolbachia* and filarial worms, the interactions with the host immune response during filarial infection and the potential for targeting *Wolbachia* in treatment and control of filariasis [2, 3, 8, 13]. Therefore, the current review will focus primarily on the host immune response to *Wolbachia* in the pathogenesis of ocular onchocerciasis.

The Role of *Wolbachia* in the Pathogenesis of Onchocerciasis and Lymphatic Filariasis

Most of the evidence supporting a role for *Wolbachia* in the pathogenesis of filarial diseases stems from posttreatment reactions in infected individuals. For example, systemic treatment with diethylcarbamazine (DEC) of onchocerciasis patients causes rapid death of microfilariae in the skin and eyes, resulting in often severe posttreatment side effects termed the Mazzotti reaction. This response, the severity of which is dependent on the number of microfilariae in the skin, is characterized by fever, tachycardia, and severe pruritus [14].

Although these earlier studies did not address the role of *Wolbachia*, the symptoms are consistent with localized responses to bacterial products. Ivermectin also targets microfilariae, has fewer side effects and has replaced DEC as the drug of choice for onchocerciasis [3]. However, posttreatment reactions also occur with ivermectin, and are associated with proinflammatory cytokines in serum, and with elevated *Wolbachia* DNA [15, 16]. Adverse posttreatment reactions also occur in lymphatic filariasis patients after treatment with the filaricidal drug DEC, *Wolbachia* DNA and occasionally intact *Wolbachia* being detected in the bloodstream [17].

Converse studies examined the effect of doxycycline on *Wolbachia* in *O. volvulus* in infected individuals. Initial studies examined adult *O. volvulus* worms in nodules recovered from individuals treated with or without doxycycline (both groups were given ivermectin, which kills microfilariae but not adult worms). Results of these treatments showed *Wolbachia* depleted from adult worms after treatment with doxycycline with the effect on embryogenesis described above [7]. Human monocytes incubated with *O. volvulus* extracts containing *Wolbachia* stimulated the production of proinflammatory and chemotactic cytokines compared with extracts from *O. volvulus* nodules in the absence of *Wolbachia* [18]. Furthermore, *Wolbachia* are required for recruitment of neutrophils to the onchocercomata (skin nodules containing *O. volvulus*), as the number of neutrophils in nodules from doxycycline-treated individuals is greatly reduced compared with untreated individuals [18].

Consistent with this finding, doxycycline treatment of individuals with lymphatic filariasis not only reduces the number of *Wolbachia*, but also reduces plasma vascular endothelial growth factor and significantly improves pathology in lymphatic filariasis, with reduced lymphedema in doxycycline-treated compared with untreated patients [19]. The onset of clinical disease manifestations also coincides with elevated levels of anti-WSP IgG in the blood of humans and rhesus monkeys [20, 21].

An additional line of evidence for a role for *Wolbachia* in the pathogenesis of onchocerciasis relates to earlier studies showing that two strains of *O. volvulus* that differ in virulence exist in West Africa based on DNA probes using a noncoding repeat sequence [22, 23]. In a recent study, the strain shown to cause more severe ocular disease had significantly higher *Wolbachia* loads compared with the second, less virulent strain, indicating a correlation between virulence and *Wolbachia* in ocular onchocerciasis [24].

Taken together, these findings strongly support the notion of an important role for *Wolbachia* in the proinflammatory response and pathogenesis of onchocerciasis and lymphatic filariasis in individuals infected with filarial nematodes.

The Role of *Wolbachia* in the Pathogenesis of Ocular Onchocerciasis

O. volvulus microfilariae released from female worms in subcutaneous nodules migrate through the skin and can invade the anterior and posterior segments of the eye. Posterior segment disease is manifested as choriouveitis and chorioretinitis that result in loss of vision and complete blindness. In the anterior segment, which has been studied in more detail, microfilariae infiltrate the cornea, where they die as a result of anti-filarial therapy or by natural attrition, and begin to degenerate, presumably releasing *Wolbachia* into the immediate environment. The host inflammatory response results in cellular infiltration, loss of corneal clarity and visual impairment [8]. Repeated infiltration in chronically infected individuals eventually leads to scar formation and complete blindness.

Given the paucity of human corneal material, the approach to understanding the pathogenesis has been to develop animal models, and a mouse model was used to investigate the role of *Wolbachia* in the pathogenesis of *O. volvulus* keratitis. In this model, filarial extracts with or without *Wolbachia* were injected directly into the corneal stroma, and keratitis is assessed by changes in corneal morphology including increased corneal thickness and haze, and by recruitment of neutrophils to the corneal stroma. Two approaches were taken: (1) injection of *O. volvulus* extracts from the patients described above who were treated with ivermectin alone (containing *Wolbachia*) or with ivermectin and doxycycline (depleted of *Wolbachia*), and (2) injection of other filarial species either containing *Wolbachia* (*B. malayi*) or that do not harbor *Wolbachia* (*A. viteae*). In both sets of experiments, neutrophil infiltration and development of corneal disease occurred when *Wolbachia* were present, whereas injection of filarial extracts without *Wolbachia* did not induce keratitis [25]. Furthermore, intracorneal injection of *Wolbachia* isolated from a mosquito cell line induced neutrophil infiltration and development of corneal haze in the absence of filarial antigens, thereby demonstrating that *Wolbachia* can directly induce keratitis [26, 27].

In human disease, *Wolbachia* arrive in the cornea within microfilariae; therefore, it was of interest to determine the fate of *Wolbachia* in this context. *B. malayi* microfilariae were recovered from infected gerbils and injected live into the corneal stroma, and *Wolbachia* were detected using immunogold labeling with antibody to the major WSP [27]. Figure 1 shows the presence of *Wolbachia* in the microfilariae in the cornea after 4 and 18 h, with neutrophils surrounding the worms and in immediate proximity to *Wolbachia*. Figure 2 shows immunogold labeling in neutrophil vacuoles surrounded by primary granules, supporting the notion that *Wolbachia* are ingested by neutrophils.

Incubation of neutrophils with *Wolbachia* stimulates release of proinflammatory and chemotactic cytokines, which contributes further to the inflammatory process in the cornea [27].

Wolbachia and Toll-Like Receptors

Toll-like receptors (TLRs) are a class of innate receptors initially identified by genomic comparison to the *Drosophila toll* gene. Toll which encodes an essential anti-microbial protein as Toll-deficient mutants succumb to extensive fungal infection [28, 29]. In mammals, the first Toll homologue to be identified was TLR4, which is critical for recognition of bacterial lipopolysaccharide [30]. Since then, at least twelve TLRs have been reported that recognize other microbial products. One of them, TLR2, forms heterodimers with TLR1 and TLR6, and its ligands include bacterial lipoproteins associated with several bacteria and fungi [31, 32].

Using a reporter cell line expressing specific human TLR and macrophages from TLR gene knockout mice, TLR2 and TLR6 were found to be the major receptors for *Wolbachia* using filarial extracts or *Wolbachia* isolated from *B. malayi* [33]. In contrast, filarial extracts from *A. viteae* which does not harbor *Wolbachia*, or from *B. malayi* isolated from gerbils treated with tetracycline did not activate TLR2 [33]. Recombinant WSP also activates TLR2 [34], indicating that this protein contains a TLR2 ligand. Earlier studies from our lab and others showed that *Wolbachia* and WSP activated TLR4 [25, 34, 35]. However, the *Wolbachia* genome sequence revealed no lipopolysaccharide synthase enzymes and therefore cannot produce endotoxin [36, 37]. TLR4 activity in these studies may have been due to trace levels of endotoxin contamination of parasite extracts. Furthermore, recent studies using preparations made under very stringent conditions did not activate TLR4 [33]. Therefore, TLR2 agonists are more likely to be surface proteins such as WSP and other cell wall components.

The cytoplasmic tail of the TLR molecule initiates cell signaling by recruiting adaptor molecules, including myeloid differentiation factor 88 (MyD88), MyD88-adaptor-like (Mal), TIR domain-containing adapter inducing IFN-β (TRIF) and TRIF-related adaptor molecule [reviewed in 38]. Consistent with signaling pathways associated with TLR2/TLR6, MyD88 and Mal were essential for *Wolbachia*-induced cytokine production, as macrophages from MyD88 and Mal gene knockout mice did not respond to *Wolbachia* or filarial extracts. No role was identified for the adaptors TRIF or TRIF-related adaptor molecule (TRAM), which are associated with TLR3 and TLR4 activation [33, 38].

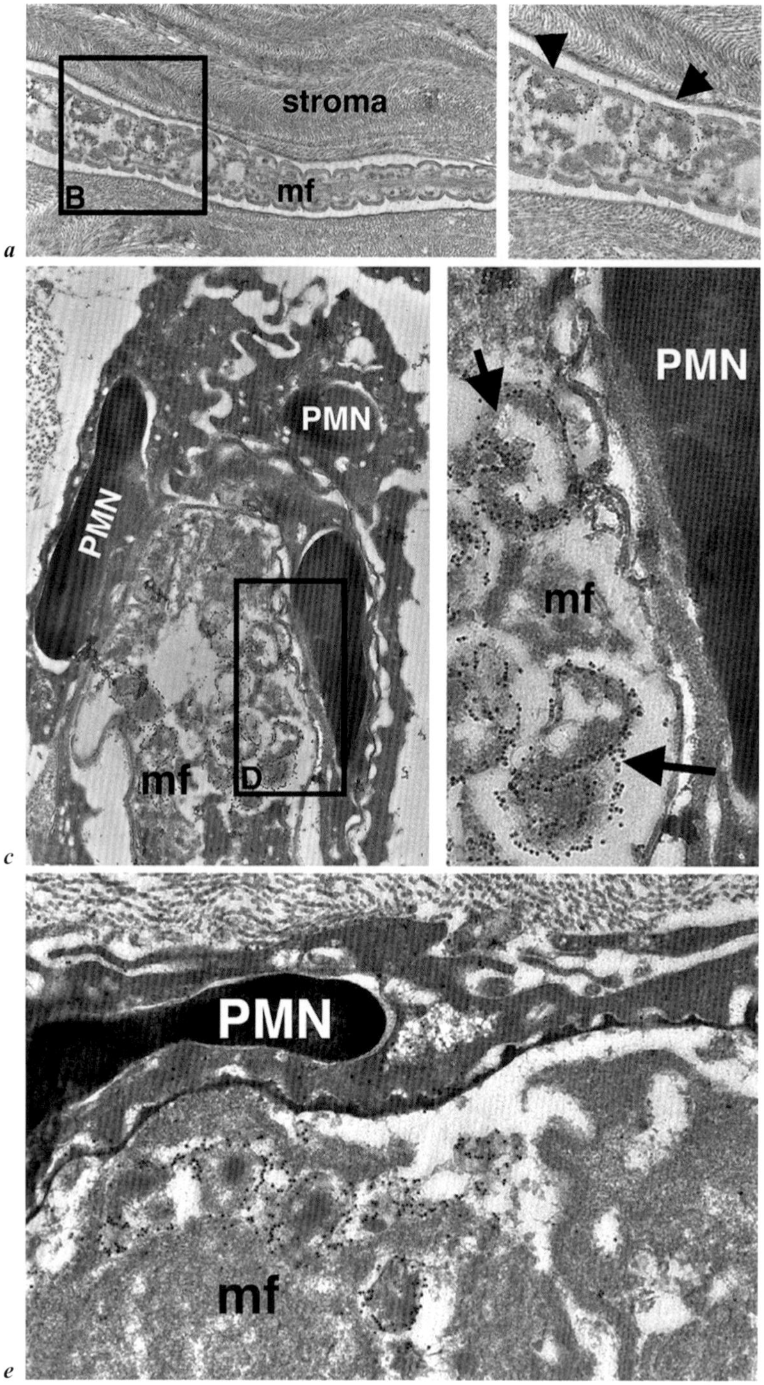

a b c d e

To determine the role of adaptor molecules in *Wolbachia*/*O. volvulus* keratitis, control and MyD88 gene knockout mice were injected with *Wolbachia* from insect cells or with *O. volvulus* extracts containing *Wolbachia*, and keratitis was assessed by neutrophil infiltration and changes in corneal structure as described above. Figure 3 shows corneal sections 18 h after injection of *Wolbachia* into the corneal stroma of MyD88$^{-/-}$ mice and wild-type littermates. In normal corneas, an intense neutrophil infiltrate is evident along with increased corneal thickening, whereas no neutrophils are detected in MyD88$^{-/-}$ mice and corneal thickness is similar to control, saline-injected mice. These findings clearly demonstrate that *Wolbachia* can induce keratitis in the absence of filarial antigens, and that MyD88 is absolutely essential for development of keratitis. Similar results were found for development of corneal haze after injection of *Wolbachia* or *O. volvulus* extracts containing *Wolbachia* [26]. As MyD88 is important for signaling by TLR2 and TLR4, single and TLR2/4 double gene knockout mice were used to examine the role of these receptors in neutrophil and macrophage infiltration to the cornea, and in development of corneal haze. In contrast to control C57BL/6 mice which develop keratitis, corneas from TLR2$^{-/-}$ and TLR2/4$^{-/-}$ mice have minimal cellular infiltration or changes in corneal clarity, whereas TLR4$^{-/-}$ corneas are similar to C57BL/6 mice [33, 39].

Taken together, these findings suggest the following sequence of events in the role of *Wolbachia*-induced corneal inflammation (fig. 4). (1) The inflammatory response to *Wolbachia* is initiated after death and degeneration of microfilariae and release of bacteria into the corneal stroma. (2) *Wolbachia* activate TLR2 and MyD88 on resident cells in the cornea, including resident fibroblasts and bone marrow-derived macrophages and dendritic cells, which produce proinflammatory and chemotactic cytokines [33, 40]. (3) These cells activate vascular endothelial cells on peripheral, limbal blood vessels to facilitate neutrophil recruitment into the avascular corneal stroma by CXC chemokines KC

Fig. 1. Proximity of neutrophils to *Wolbachia* in the nematode hypodermis. C57BL/6 mice were injected into the corneal stroma with microfilariae, corneas were removed after 4 or 18 h and thin sections were immunostained with anti-(WSP) and visualized with IgG conjugated to 15-nm gold particles. Sections were counterstained with uranyl acetate and lead citrate, and examined by electron microscopy. ***a***, ***b*** 4 h after injection. WSP was clearly detected inside microfilariae in the corneal stroma (arrows). mf = Microfilariae. ***c–e*** 18 h after injection, microfilariae containing *Wolbachia* were surrounded by neutrophils (PMN). WSP labeled with gold particles (arrows) are present in the microfilariae adjacent to the neutrophils in either unimmunized (***c***) or immunized (***e***) mice. Magnifications: ×4,800 (***a***); ×8,400 (***b***); ×5,300 (***c***); ×16,000 (***d***); ×14,500 (***e***). Reprinted with permission from Gillette-Ferguson et al. [27].

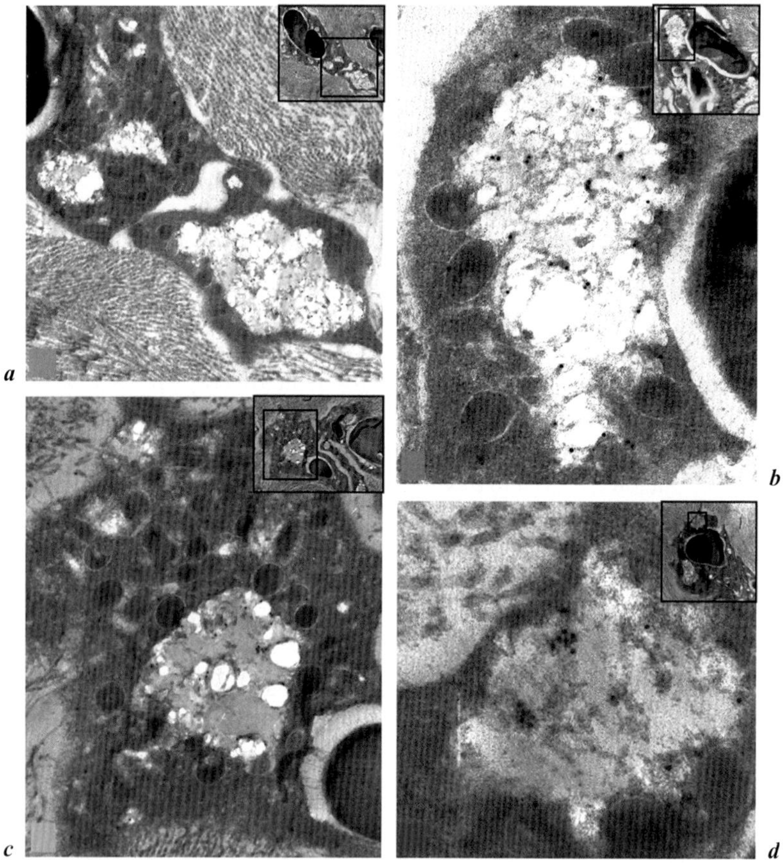

Fig. 2. *Wolbachia* in neutrophil vacuoles. Immuno-electron microscopy of neutrophils 18 h after injection of microfilariae. Immuno-gold particles specific for WSP were prominent in neutrophil vacuoles of both immunized (***a***, ***b***) and unimmunized (***c–d***) mice. Magnifications: ×11,400 (***a***); ×45,000 (***b***); ×24,000 (***c***); ×67,500 (***d***). Reprinted with permission from Gillette-Ferguson et al. [27].

and MIP-2, and CXCR2 receptor on the neutrophils [41]. Neutrophils migrate through the stromal matrix to the site of microfilarial degradation and release of *Wolbachia.* As neutrophils also express TLR2 and MyD88 [26, 39], (4) neutrophils ingest *Wolbachia* and produce proinflammatory and chemotactic cytokines [27], which (5) stimulate further neutrophil infiltration. (6) Neutrophil degranulation and secretion of cytotoxic products such as nitric oxide, myeloperoxidase and oxygen radicals have a cytotoxic effect on resident cells in the cornea, including fibroblasts and corneal endothelium, resulting in corneal edema and further loss of corneal clarity.

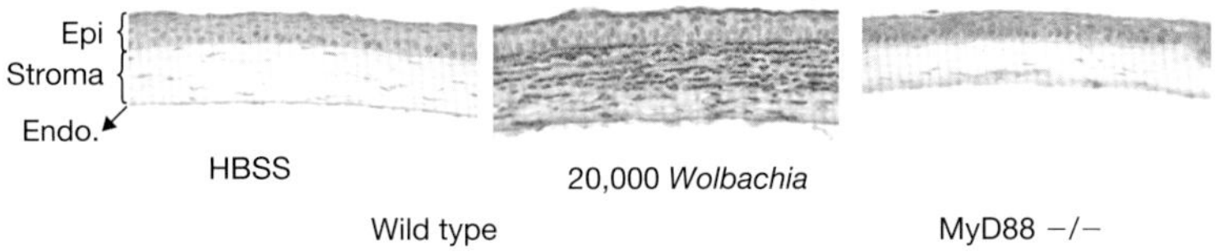

Fig. 3. Neutrophils in the corneal stroma of wild-type and MyD88$^{-/-}$ mice after injection of *Wolbachia*. *Wolbachia* were injected into the corneal stroma of MyD88$^{-/-}$ mice and wild-type littermates. After 18 h, mice were sacrificed and corneas stained for neutrophils using MAb NIMP-R14. Representative sections of wild-type and MyD88$^{-/-}$ mice injected with 4 μl saline (HBSS) or with 20,000 bacteria. Reprinted with permission from Gillette-Ferguson et al. [26]

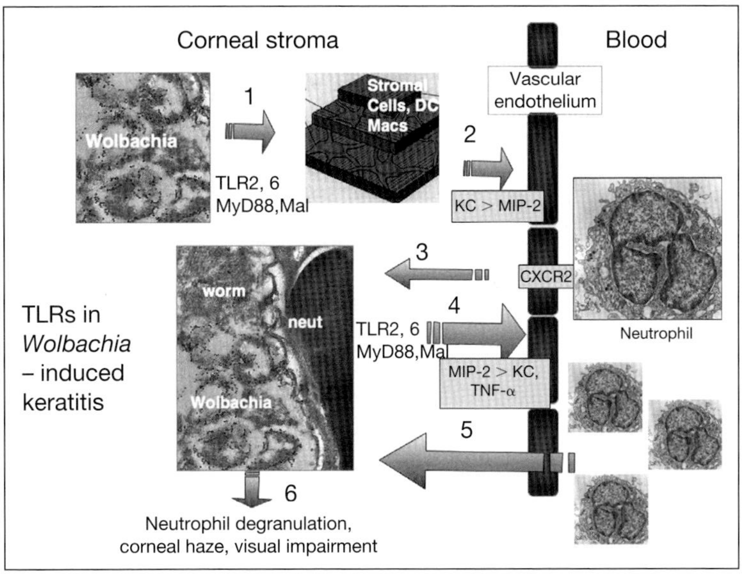

Fig. 4. Proposed sequence of events in innate immune responses underlying *O. volvulus*-induced keratitis (see text for description).

The Role of TLR and Adaptor Molecules in the Adaptive Immune Response in *Wolbachia*- and *O. volvulus*-Induced Keratitis

The proposed sequence of events is in relation to the acute response to invading microfilariae containing *Wolbachia*; however, in chronically infected, untreated individuals, there is also an ongoing adaptive immune response, with repeated invasion of microfilariae into the corneal stroma, and consistent worm degeneration and release of *Wolbachia* [2, 8]. Dendritic cells represent a bridge between the innate and adaptive arms of the immune response, as they travel from the site of initial antigen retrieval to draining lymph nodes where they present antigens to T cells. Activation of dendritic cells by WSP or by filarial extracts containing *Wolbachia* requires expression of TLR2 and MyD88 [34, 40]. In addition, TLR2, but not TLR4 mediates T helper cell type 1 (Th1) responses such as IFN-γ production, but not Th2-associated responses including serum IgE and IgG_1, IL-5 production and eosinophil migration to the corneal stroma [40]. Interestingly, although antibody and immune complexes are important in neutrophil recruitment to the cornea [42, 43], TLR2 appears to be the dominant signal for neutrophil recruitment to the cornea because, even though there are high levels of circulating anti-filarial antibody, neutrophil recruitment is significantly reduced in TLR2 gene-deficient mice [40]. Also of interest is that *Wolbachia*-mediated immune tolerance is TLR2/MyD88 dependent, and may have an important role in suppression of the host immune response during chronic infection and high levels of parasitemia [44].

In contrast to filarial nematodes that do harbor *Wolbachia*, a secretory product from *A. viteae* (ES-62) which does not harbor *Wolbachia* downmodulates the host response in filarial infections by activating TLR4 [45], implicating an immunomodulatory role for helminth antigens, the mechanism of suppression has yet to be determined.

Conclusion

Findings from studies on infected individuals and animal models demonstrate that endosymbiotic *Wolbachia* have a profound effect on the pathogenesis of filarial disease, although the effect of *Wolbachia* and TLRs depends on the experimental system being studied. For corneal disease in ocular onchocerciasis, it is clear that the *Wolbachia*-induced TLR2 activation mediates at least the initial stages of the inflammatory response in the cornea. Given that antibiotic treatment eliminates *Wolbachia* from worms in infected individuals, the data reviewed here indicate that this treatment will also limit the severity of corneal disease. Future studies will continue to explore the relationship between *Wolbachia*, TLR activation and disease manifestations in infected populations.

Acknowledgements

This work was supported by NIH grants EY10320, EY11373 (E.P.), and by the Research to Prevent Blindness Foundation and the Ohio Lions Eye Research Foundation.

E.P. is a recipient of an RPB senior investigator award. Studies were also supported by K08 AI054652 (A.G.H.) and DA1024/1-1 from the German Research Foundation (K.D.).

References

1 Taylor MJ, Hoerauf A: *Wolbachia* bacteria of filarial nematodes. Parasitol Today 1999;15: 437–442.

2 Brattig NW: Pathogenesis and host responses in human onchocerciasis: impact of *Onchocerca filariae* and *Wolbachia* endobacteria. Microbes Infect 2004;6:113–128.

3 Taylor MJ, Bandi C, Hoerauf A: *Wolbachia* bacterial endosymbionts of filarial nematodes. Adv Parasitol 2005;60:245–284.

4 McGarry HF, Pfarr K, Egerton G, Hoerauf A, Akue JP, Enyong P, Wanji S, Klager SL, Bianco AE, Beeching NJ, Taylor MJ: Evidence against *Wolbachia* symbiosis in *Loa loa*. Filaria J 2003;2:9.

5 Casiraghi M, Bain O, Guerrero R, Martin C, Pocacqua V, Gardner SL, Franceschi A, Bandi C: Mapping the presence of *Wolbachia pipientis* on the phylogeny of filarial nematodes: evidence for symbiont loss during evolution. Int J Parasitol 2004;34:191–203.

6 Casiraghi M, Bordenstein SR, Baldo L, Lo N, Beninati T, Wernegreen JJ, Werren JH, Bandi C: Phylogeny of *Wolbachia pipientis* based on gltA, groEL and ftsZ gene sequences: clustering of arthropod and nematode symbionts in the F supergroup, and evidence for further diversity in the *Wolbachia* tree. Microbiology 2005;151:4015–4022.

7 Hoerauf A, Mand S, Adjei O, Fleischer B, Buttner DW: Depletion of *Wolbachia* endobacteria in *Onchocerca volvulus* by doxycycline and microfilaridermia after ivermectin treatment. Lancet 2001;357:1415–1416.

8 Hoerauf A, Buttner DW, Adjei O, Pearlman E: Onchocerciasis. BMJ 2003;326:207–210.

9 Hoerauf A, Mand S, Volkmann L, Buttner M, Marfo-Debrekyei Y, Taylor M, Adjei O, Buttner DW: Doxycycline in the treatment of human onchocerciasis: kinetics of *Wolbachia* endobacteria reduction and of inhibition of embryogenesis in female *Onchocerca* worms. Microbes Infect 2003;5: 261–273.

10 Hoerauf A, Volkmann L, Hamelmann C, Adjei O, Autenrieth IB, Fleischer B, Buttner DW: Endosymbiotic bacteria in worms as targets for a novel chemotherapy in filariasis. Lancet 2000;355: 1242–1243.

11 Taylor MJ, Makunde WH, McGarry HF, Turner JD, Mand S, Hoerauf A: Macrofilaricidal activity after doxycycline treatment of *Wuchereria bancrofti*: a double-blind, randomised placebo-controlled trial. Lancet 2005;365:2116–2121.

12 McGarry HF, Egerton GL, Taylor MJ: Population dynamics of *Wolbachia* bacterial endosymbionts in *Brugia malayi*. Mol Biochem Parasitol 2004;135:57–67.

13 Hoerauf A: New strategies to combat filariasis. Expert Rev Anti Infect Ther 2006;4:211–222.

14 Francis H, Awadzi K, Ottesen EA: The Mazzotti reaction following treatment of onchocerciasis with diethylcarbamazine: clinical severity as a function of infection intensity. Am J Trop Med Hyg 1985;34:529–536.

15 Awadzi K, Opoku NO, Addy ET, Quartey BT: The chemotherapy of onchocerciasis. XIX: the clinical and laboratory tolerance of high dose ivermectin. Trop Med Parasitol 1995;46:131–137.

16 Keiser PB, Reynolds SM, Awadzi K, Ottesen EA, Taylor MJ, Nutman TB: Bacterial endosymbionts of *Onchocerca volvulus* in the pathogenesis of posttreatment reactions. J Infect Dis 2002;185:805–811.

17 Cross HF, Haarbrink M, Egerton G, Yazdanbakhsh M, Taylor MJ: Severe reactions to filarial chemotherapy and release of *Wolbachia* endosymbionts into blood. Lancet 2001;358:1873–1875.

18 Brattig NW, Buttner DW, Hoerauf A: Neutrophil accumulation around *Onchocerca* worms and chemotaxis of neutrophils are dependent on *Wolbachia* endobacteria. Microbes Infect 2001;3: 439–446.

19 Debrah AY, Mand S, Specht S, Marfo-Debrekyei Y, Batsa L, Pfarr K, Larbi J, Lawson B, Taylor M, Adjei O, Hoerauf A: Doxycycline reduces plasma VEGF-C/sVEGFR-3 and improves pathology in lymphatic filariasis. PLoS Pathog 2006;2:e92.

20 Lammie PJ, Cuenco KT, Punkosdy GA: The pathogenesis of filarial lymphedema: is it the worm or is it the host? Ann NY Acad Sci 2002;979:131–142; discussion 188–196.

21 Punkosdy GA, Addiss DG, Lammie PJ: Characterization of antibody responses to *Wolbachia* surface protein in humans with lymphatic filariasis. Infect Immun 2003;71:5104–5114.

22 Erttmann KD, Unnasch TR, Greene BM, Albiez EJ, Boateng J, Denke AM, Ferraroni JJ, Karam M, Schulz-Key H, Williams PN: A DNA sequence specific for forest form *Onchocerca volvulus*. Nature 1987;327:415–417.

23 Zimmerman PA, Dadzie KY, De Sole G, Remme J, Alley ES, Unnasch TR: *Onchocerca volvulus* DNA probe classification correlates with epidemiologic patterns of blindness. J Infect Dis 1992;165:964–968.

24 Higazi TB, Filiano A, Katholi CR, Dadzie Y, Remme JH, Unnasch TR: Wolbachia endosymbiont levels in severe and mild strains of *Onchocerca volvulus*. Mol Biochem Parasitol 2005;141: 109–112.

25 Saint Andre A, Blackwell NM, Hall LR, Hoerauf A, Brattig NW, Volkmann L, Taylor MJ, Ford L, Hise AG, Lass JH, Diaconu E, Pearlman E: The role of endosymbiotic *Wolbachia* bacteria in the pathogenesis of river blindness. Science 2002;295:1892–1895.

26 Gillette-Ferguson I, Hise AG, Sun Y, Diaconu E, McGarry HF, Taylor MJ, Pearlman E: *Wolbachia*- and *Onchocerca volvulus*-induced keratitis (river blindness) is dependent on myeloid differentiation factor 88. Infect Immun 2006;74:2442–2445.

27 Gillette-Ferguson I, Hise AG, McGarry HF, Turner J, Esposito A, Sun Y, Diaconu E, Taylor MJ, Pearlman E: Wolbachia-induced neutrophil activation in a mouse model of ocular onchocerciasis (river blindness). Infect Immun 2004;72:5687–5692.

28 Lemaitre B: The road to Toll. Nat Rev Immunol 2004;4:521–527.

29 Lemaitre B, Nicolas E, Michaut L, Reichhart JM, Hoffmann JA: The dorsoventral regulatory gene cassette spatzle/Toll/cactus controls the potent antifungal response in Drosophila adults. Cell 1996;86:973–983.

30 Medzhitov R, Preston-Hurlburt P, Janeway CA Jr: A human homologue of the Drosophila Toll protein signals activation of adaptive immunity. Nature 1997;388:394–397.

31 Akira S: TLR signaling. Curr Top Microbiol Immunol 2006;311:1–16.

32 Akira S, Uematsu S, Takeuchi O: Pathogen recognition and innate immunity. Cell 2006;124: 783–801.

33 Hise AG, Daehnel K, Gillette-Ferguson I, Cho E, McGarry HF, Taylor M, Golenbock DT, Fitzgerald KA, Kazura JW, Pearlman E: Innate immune responses to endosymbiotic *Wolbachia* bacteria in *Brugia malayi* and *Onchocerca volvulus* are dependent on TLR2, TLR6, MyD88 and Mal, but not TLR4, TRIF or TRAM. J Immunol 2007;178:1068–1076.

34 Brattig NW, Bazzocchi C, Kirschning CJ, Reiling N, Buttner DW, Ceciliani F, Geisinger F, Hochrein H, Ernst M, Wagner H, Bandi C, Hoerauf A: The major surface protein of *Wolbachia* endosymbionts in filarial nematodes elicits immune responses through TLR2 and TLR4. J Immunol 2004;173:437–445.

35 Taylor MJ, Cross HF, Bilo K: Inflammatory responses induced by the filarial nematode *Brugia malayi* are mediated by lipopolysaccharide-like activity from endosymbiotic *Wolbachia* bacteria. J Exp Med 2000;191:1429–1436.

36 Blaxter M, Daub J, Guiliano D, Parkinson J, Whitton C: The *Brugia malayi* genome project: expressed sequence tags and gene discovery. Trans R Soc Trop Med Hyg 2002;96:7–17.

37 Wu M, Sun LV, Vamathevan J, Riegler M, Deboy R, Brownlie JC, McGraw EA, Martin W, Esser C, Ahmadinejad N, Wiegand C, Madupu R, Beanan MJ, Brinkac LM, Daugherty SC, Durkin AS, Kolonay JF, Nelson WC, Mohamoud Y, Lee P, Berry K, Young MB, Utterback T, Weidman J, Nierman WC, Paulsen IT, Nelson KE, Tettelin H, O'Neill SL, Eisen JA: Phylogenomics of the reproductive parasite *Wolbachia pipientis* wMel: a streamlined genome overrun by mobile genetic elements. PLoS Biol 2004;2:E69.

38 Vogel SN, Fitzgerald KA, Fenton MJ: TLRs: differential adapter utilization by toll-like receptors mediates TLR-specific patterns of gene expression. Mol Interv 2003;3:466–477.

39 Gillette-Ferguson I, Daehnel K, Hise AG, Sun Y, Carlson E, Diaconu E, Taylor MJ, Pearlman E: TLR2 expression on resident corneal cells or on bone marrow derived cells is sufficient to mediate neutrophil migration to the corneal stroma and development of *Onchocerca volvulus/Wolbachia keratitis* (river blindness), submitted.
40 Daehnel K, Gillette-Ferguson I, Hise AG, Diaconu E, Harling MJ, Heinzel FP, Pearlman E: Toll Like Receptor (TLR) 2 is essential for development of T helper (Th) type 1, but not Th2-associated responses to filarial antigens, submitted.
41 Hall LR, Diaconu E, Patel R, Pearlman E: CXC chemokine receptor 2 but not C-C chemokine receptor 1 expression is essential for neutrophil recruitment to the cornea in helminth-mediated keratitis (river blindness). J Immunol 2001;166:4035–4041.
42 Hall LR, Diaconu E, Pearlman E: A dominant role for Fc gamma receptors in antibody-dependent corneal inflammation. J Immunol 2001;167:919–925.
43 Hall LR, Lass JH, Diaconu E, Strine ER, Pearlman E: An essential role for antibody in neutrophil and eosinophil recruitment to the cornea: B cell-deficient (microMT) mice fail to develop Th2-dependent, helminth-mediated keratitis. J Immunol 1999;163:4970–4975.
44 Turner JD, Langley RS, Johnston KL, Egerton G, Wanji S, Taylor MJ: *Wolbachia* endosymbiotic bacteria of *Brugia malayi* mediate macrophage tolerance to TLR- and CD40-specific stimuli in a MyD88/TLR2-dependent manner. J Immunol 2006;177:1240–1249.
45 Goodridge HS, Marshall FA, Else KJ, Houston KM, Egan C, Al-Riyami L, Liew FY, Harnett W, Harnett MM: Immunomodulation via novel use of TLR4 by the filarial nematode phosphorylcholine-containing secreted product, ES-62. J Immunol 2005;174:284–293.

Eric Pearlman, PhD
Department of Ophthalmology, Case Western Reserve University
10900 Euclid Ave
Cleveland, OH 44106-7286 (USA)
Tel. +1 216 368 1856, Fax +1 216 368 4825, E-Mail Eric.Pearlman@case.edu

Author Index

Subject Index